Chapter 1

A 'FINAL GESTURE' AT OLDHAM

The invasion of Russia in June 1941 had taken the bulk of the German bomber force eastwards, and from the end of August 1942 even small scale and sporadic air raids on the North West of England ceased.

For Civil Defence purposes Region 10, the North West, ran southwards from the Scottish border to include the counties of Cumberland, Westmorland, Lancashire, Cheshire and North West Derbyshire. By August 1944 the Ministry of Home Security felt able to issue instructions for the easing of all Civil Defence duties in parts of the country away from London and the South East of England. This was much to the relief of part-time workers in the Civil Defence services, for whom the strains of long hours of war work as well as routine Civil Defence duties were not to be borne happily for years on end. They were still, after the easing, required to stay enrolled to maintain their equipment and to turn out for duty in the event of an air raid alert or of enemy action.

The regular manning of Civil Defence posts, however, became little more than symbolic and the Civil Defence services within the region went into severe decline.

This was in sharp contrast to the period during and just after the Blitz raids of 1940 and 1941. At the peak, 9,342 part-time men and 3,592 part-time women were enrolled and active in the various Civil Defence services of the region. Additionally, the great firebomb raids of December 1940 on London and Manchester gave rise to the formation of a National Fireguard Service, formed to stand guard over the roofs of buildings and there to deal with incendiary bombs before these were able to start serious fires. This service had peaked in the Region at 120 full time staff and 240,000 part timers. All service was compulsory under the Emergency Regulations of the time. The Fireguard Services, too, were "stood down" in mid-September.

Whilst the part-timers welcomed their increased leisure time, the 609 men and 238 women (exclusive of ambulance workers) employed full-time in the Civil Defence Services of the Region were not so happy with their lot. All except a small cadre were demobilised and transferred to other work from 15th October 1944. As in most redundancy situations, there was some consequent dissatisfaction and this was reflected in a letter from "a full time warden" published in the Daily Dispatch of 11th September:-

"Sir - Permit me to protest against the attitude now being taken up by the authorities against full time Civil Defence workers. To men who have not only served five years full time, but also more than a year voluntary before the war, the final gesture is at the same time parsimonious and undignified.

To come out into the open and give the whole organisation a fixed notice would not have been resented, but the press or the BBC has each day something to say about the disposal of these men, as if they were something to be got rid of as soon as possible, a sore on the municipal body.

The public looked [illegible] wardens when the raids were on and called out for their help, but as soon as the raids were over, talked as if they were keeping a lot of pensioners."

The relaxed public attitude to the possibility of further air raids on the Region was highlighted at a special meeting of the Manchester War Emergency Committee in November 1944, when it was stated that the number of applications from people wishing to return Anderson and Morrison home air raid shelters (the first an outside and the second an inside type) to the authority was increasing weekly and that storage accommodation was consequently becoming inadequate. The Committee, therefore, issued a warning that it was an offence under the Emergency Regulations then in force to remove a shelter and that no more could be collected from the public until the

BLACK-OUT OFF

From Sunday week, Sept. 17

FIRE GUARD

Ends Tuesday in most areas

HOME GUARD

No compulsion from Monday

C.D. CUTS

Workers switch to Forces or factories

Newspaper headlines on the morning of 7th September 1944 announced a general relaxation of Civil Defence duties. But the danger of enemy attack was far from over

Ministry issued instructions to collect the shelters in.

In Manchester, under the new stand down arrangements, many Civil Defence control centres were closed and their function transferred to the Regional Control Centre at Preston. This would in future deal with calls for centralised Civil Defence services through the organisation of mobile columns which could be despatched to particular incidents throughout the Region.

Mr (later Lord) Hartley Shawcross was to remain the Regional Commissioner for Civil Defence; he issued the detailed directions for, and implemented, the stand down.

During the long period of enemy inactivity over the Region, some Manchester members of Civil Defence had been temporarily transferred to areas still subject to attack. The Manchester Evening News reported on 3rd June 1944 that local Civil Defence members attended the funeral of Thomas Edwards of New Moston, a member of the Civil Defence Reserve Service who was killed on active duty in a South West town (almost certainly Plymouth) when a blitzed wall fell on him. He had been temporarily transferred in January 1944, at the beginning of what became known as the "Baby Blitz".

Official complacency regarding the need for adequate civil defence in the North was to receive a rude shock which, but for the vagary of chance, might have had much worse consequences than it did.

Wilfred Swann had spent the afternoon of Saturday 23rd December 1944 at the Tatler News Theatre off Oxford Road in Manchester. Amongst the films he watched there was a documentary entitled "Story of the V-1". The sound of the pulse jet motor of the V-1 weapon from the film soundtrack, once graphically described as "the hideous noise, coarser, louder, more blatant than that of a regular aircraft engine," was still fresh in his mind that same night. It was cold, dark and overcast at a quarter past five, when he was wakened by the undulating wail of the Oldham air raid siren. He had already started his Christmas break from employment at Ferranti's Moston works, as it was Sunday morning and Christmas Eve. He was, however, a part-time National Fire Service fireman and in that capacity bound to turn out for duty on the alert.

He was quickly out of bed and into his uniform, but before setting out for his post he needed a quick visit to his privy in the back garden. Here outside events overtook him; still closeted, he recognised the approaching sound immediately, one not dissimilar to that of a motorcycle engine. It was the same sound exactly as that on the film soundtrack - that of a flying bomb. It got rapidly louder as he rushed towards his back door, shouting to his wife and family to get under the kitchen table. Just as he went in through the door the noise stopped. A brief silence was broken a few seconds later by a flash and a muffled explosion followed by the sound of breaking glass. The plate glass windows of the local Co-operative shop were blown outwards into the street by the blast of the detonation.

V-1 damage in Abbey Hills Road, Oldham. A three-storey house at the end of a block of twelve on the north side of the road

The bomb had fallen on Abbey Hills Road about three-quarters of a mile to the east of the Swanns' home in Orme Street. Owing to the contours of the ground, the Orme Street area had been relatively protected from the blast effect.

For a short while there was some alarm and speculation as to what would follow and then came the sound of a fire engine on Park Road heading towards Abbey Hills Road. Wilfred Swann decided to follow this to the incident rather than to go to his post, reasoning that his unit would be directed there to assist. The news quickly passed around that the site of the explosion was in Abbey Hills Road. As he neared it the signs of damage became steadily more severe. At Skipton Street on the right was his aunt's house, only two hundred yards from the impact point. The house had its front door blown in and much of the ceiling was down. A layer of black powder, a mixture of soot and dust, had coated everything including the people in the vicinity. Aunty was not seriously injured and had gone to friends in Kearsley Street. It was later found that the whole roof of the house had been lifted by the blast and moved seven inches.

The centre of the explosion, close to the junction with Warren Lane, was a scene of devastation on both sides of Abbey Hills Road.

He joined a number of Civil Defence rescuers already at work and stayed until finally relieved at four in the afternoon. His services were particularly valuable as he knew most of the people living in the worst affected part of the road. The Swann family were lucky not to have been victims of the bomb themselves, since they had previously considered an offer of one of the houses now severely damaged, but turned it down because of a series of steps at the back which made it unsuitable for Mother-in-law's rheumatic condition.

The dead, and those too badly injured to walk, when they were released from the debris, were passed on stretchers over the rubble to waiting ambulances, and the less seriously injured treated in a first aid post set up in the damaged Congregational chapel on Nether Hey Street close by.

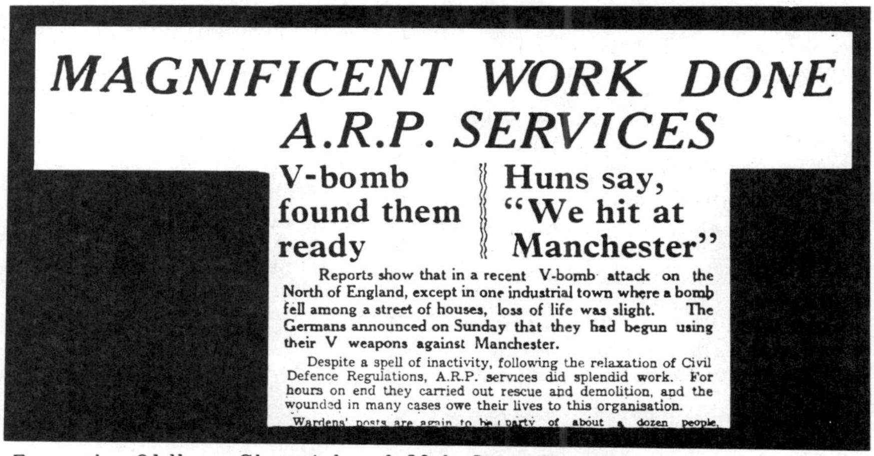

MAGNIFICENT WORK DONE A.R.P. SERVICES

V-bomb found them ready

Huns say, "We hit at Manchester"

Reports show that in a recent V-bomb attack on the North of England, except in one industrial town where a bomb fell among a street of houses, loss of life was slight. The Germans announced on Sunday that they had begun using their V weapons against Manchester.

Despite a spell of inactivity, following the relaxation of Civil Defence Regulations, A.R.P. services did splendid work. For hours on end they carried out rescue and demolition, and the wounded in many cases owe their lives to this organisation.

From the Oldham Chronicle of 30th December 1944

As the work proceeded there were periodic calls for silence whenever signs or sounds indicated that there may have been survivors under the wreckage. In one house, after such a call, a woman and a small boy who had been sheltering in a cellar were rescued. One horrific find was a dead dog skinned by the blast, which was at first taken for the mutilated remains of a human baby.

In another house a woman victim was found dead in bed, killed by a falling roof joist which had struck her head. The ground floors of some of the houses had collapsed into the cellars and the first floors to ground floor level. In another severely damaged house, Mr and Mrs Kirkland were entertaining a wedding party of about a dozen people who had stayed overnight. Four of them were killed, an eight-year-old girl and four women being injured. The bride and groom, fortunately, were staying in another house a short distance away.

Nearby, Mrs Bickerdike had two lodgers and three little evacuees in her house. The children were being visited by their parents, Mr and Mrs Hyman of Streatham, London, for the Christmas Holiday. Mr Hyman wakened to the sound of the approaching bomb and immediately rushed upstairs in the darkness, knowing what the sound signified and wishing to protect his children. Before he had even reached their room the explosion occurred. Mrs Hyman was the only one of the family to be injured. She and her husband had been "bombed out", as the term then was for being made homeless through enemy action, three times.

The two sons of Mr and Mrs Paul Travis, Keith and Norman, were killed; the latter was home on Christmas leave from the Royal Navy. Mrs Travis, their daughter, Pauline, and a young friend were taken to Oldham Royal Infirmary, the hospital to which all the seriously injured from the incident were taken.

Two baby boys, one of them a year old and the other only six months, were the youngest of those who lost their lives. Clifford Barker, aged nine, and Kathleen, aged six, a brother and sister, were also killed. The oldest victim was Mrs Lucy Thornton, aged seventy. Of those taken to hospital, the youngest patient was a six-weeks-old baby.

Some people, left with only the clothes they stood up in, were later issued with emergency clothing ration coupons.

Guests staying over for the Christmas period increased the normal population considerably and were not easily identified. Similarly, local residents away from home had to be traced. This made the job of accounting for all the residents much more difficult for the rescuers.

An incident enquiry centre was set up, staffed by wardens and by members of the Women's Voluntary Service. A rest centre was also organised for the homeless, pending their billeting with friends and neighbours.

When rescue operations were at their height, shortly after 8.00am, Hartley Shawcross visited the scene; he came back again on Christmas Day and afterwards visited the injured in hospital.

Rescue reinforcements, a Regional Civil Defence Column, were sent by Regional Control at Preston and provided much-needed relief for the local services. First aid and temporary repairs to the damaged property were quickly started and as much as possible was made weatherproof. This part of the work was undertaken, under the supervision of the Oldham Borough Engineer, Mr G E Hardy, mainly by local builders and contractors. Help, both in the form of labour and of materials, was sent in by surrounding districts, particularly by Manchester.

There were no obvious signs of remains of the bomb around the damaged area and there was no crater since a direct hit had been made on dwelling houses. The fuse system of a flying bomb was designed for sensitivity and rapid detonation to inflict the maximum amount of blast damage. The explosion tended to blow most of the missile into small fragments which were spread over a wide area.

In total, 32 people lost their lives in this incident and 49 were more or less seriously

Civil Defence workers search for victims at Abbey Hills Road

injured; 38 of these were kept in hospital. Of one woman no trace was ever found and a Coroner's enquiry subsequently concluded that she was a victim of the bomb and must be counted amongst the dead.

The Borough Surveyor classified the damage done to buildings in accordance with the official Bomb Census categories of the time and prepared the table reproduced here. In addition, approximately 1,100 houses in the town suffered some damage.

One badly damaged block of twelve houses on the north side of Abbey Hills Road consisted of a terrace of ten two-storey houses with a three-storey house at each end. The remainder of the damaged property was of two-storey construction (with some cellars) and was generally of 9" exterior brickwork with $4\frac{1}{2}$" party walls. All the party walls, however, contained substantial chimney breasts and fortunately these tended to buttress the structures.

This was the last full-scale "major" incident - that is, the last in which 20 or more people were killed - in the whole flying bomb campaign against Britain. However, other casualties were to occur, and more damage was to be done in Northern England that night. A colleague of Wilfred Swann's at Ferranti lived between Holmfirth and Oldham. He heard six separate flying bombs.

At Latimer Street, Oldham, a 17-year-old girl shared a bedroom with her sister. Waking, and hearing the unfamiliar sound of the pulse jet, she suggested that the noise was that of a "'plane in trouble". Hearing the explosion, but unaware that an air raid alert was in progress, they speculated that the 'plane must have hit one of the many mill chimneys in the district.

There was not much traffic on the roads during the war, particularly at night, so that for weeks afterwards many people were apprehensive of any noise from a loud motor vehicle, particularly a motorcycle, in the hours of darkness. The noise caught the attention for fear it should turn out to be another of the destructive, dangerous "buzzbombs".

A bundle of leaflets was released by the robot as it commenced its dive and these were found in Lees Cemetery on Sunny Bank Road, three-quarters of a mile due east of Abbey Hills Road. They were taken by the police, less a few that escaped into private hands!

	CATEGORY OF DAMAGE		
	A	B	C
	Destroyed	Very Severely Damaged	Severely Damaged
North Side Abbey Hills Road	9	9	24
South Side Abbey Hills Road	9	3	20
South Side Warren Street		6	9
Nether Hey Street			10
Opposite Congregational Church			10
	18	18	73

Even today, anyone standing in Abbey Hills Road, close to the junction with Warren Lane, can see distinctly the changed style of the houses which replaced the destroyed property and contrast it with the older buildings which were repaired after the explosion.

Chapter 2

A REVERSAL OF FORTUNE

At a quarter past three on the morning of 20th June 1940 the first air raid alert was sounded in the Manchester area. It lasted for 33 minutes, but there was no hostile activity. It heralded, however, a year of blitz. On 29th July, bombs fell in Salford and damage was done for the first time, this in the Ordsall Lane area. A leaflet raid on Salford followed on 8th August and from then on a succession of raids came to a climax on the nights of 22nd/23rd and 23rd/24th December 1940. The worst raids of 1941 were those of the nights of 11th March and 1st/2nd June, but after that, activity fell off sharply. Even so, by the end of 1942 there had been 329 air raid alerts sounded in Manchester.

At 8.25 on the morning of 27th July 1942, a single enemy aircraft flying low over the rooftops dropped a stick of four bombs in Palmerston Street, Hillkirk Street and Russell Street, Beswick, near the city centre. Three people were killed, seven seriously injured and twenty-one slightly injured. One of the bombs failed to explode and was dealt with by a bomb disposal unit. There were five alerts in August and bombs were dropped at Southport, Middleton, Widnes and near Warrington, but enemy activity over the region ceased after that slight flurry.

It was not at first apparent that Germany's war with Russia would be a long struggle and that the lull in air raiding would continue. Every opportunity was taken to increase Civil Defence readiness; as more equipment became available the level of training improved.

The regular testing of the air raid sirens on the first Monday of each month (All Clear - Warning - All Clear: continuous wail - undulating wail - continuous wail) were repeated reminders of the threat of the bombers' return. What was not foreseen was the sinister form the return would eventually take.

In a speech which he made at Danzig on 19th September 1939, Adolf Hitler referred to the development by Germany of secret weapons against which his enemies would be powerless. Unless this was just an idle threat, it was likely that such weapons would include long-range ones capable of striking at targets in Britain. The German military had a traditional liking for long-range artillery, a classic example of which was the so-called Paris gun (in fact there were four of them) of World War I, able to shell that city at a range of nearly 70 miles.

The German occupation of the southern coastline of the English Channel in 1940 made such bombardment of Britain by long-range weapons the more likely. Indeed, two very long-range railway-mounted guns, 21cm calibre K12s based on the Paris gun design, were moved up to the Channel coast in the late summer of 1940. These each weighed 297 tons (302 tonnes), fired a shell weighing 236.5 lbs (107.5 kg) and had a maximum range of 71.5 miles (115 Km). One of the shells fired was thrown almost to Chatham, but the barrel wear rate was so great that only occasional use was made of these monsters.

An identifiable need had arisen for the German armaments industry to replace both the absent strategic bombers and the ponderous long-range guns with weapons able to strike cheaply and effectively against targets in Britain.

In the event, two such secret weapons were developed and used operationally. They were given the generic names of Vergeltungswaffe (Revenge Weapon) by the German propaganda media. Vergeltungswaffe 1 (V-1) was a small jet-propelled pilotless aeroplane carrying an explosive payload of 850 Kg (1,870 lbs) and Vergeltungswaffe 2 (V-2) a liquid fuelled rocket with a 1 tonne warhead.

The Heinkel He177 bomber

Hitler, forever affronted by the increasingly heavy British (and, from 1942, American) air attacks on Germany, repeatedly goaded his Luftwaffe chief, Goering, to mount retaliatory raids on Britain, when it would have been politic to have reserved their limited bomber strength.

German long-range 21cm Kanone 12, used to bombard South East England from 1940

The "Air Defence of Great Britain" forces, whose responsibility was the protection of the United Kingdom from enemy air attack, were able to deal very punishingly with the raiders. Nevertheless, the Luftwaffe persisted and opened a further series of attacks on 21st January 1944. By that time the defenders had truly got the measure of the raiders and shot down some 135 of them between that date and the end of the series in May. This number amounted to some 6% of the raider sorties flown. Inadequate crew training contributed to substantial further Luftwaffe losses in accidents. Some of the raids were directed at South Coast ports from which the Allied troops would embark for the impending D-Day invasion. The losses incurred and the Luftwaffe's inability to replace these left a very depleted German bomber force to face the landings, which followed a bare month after the series of raids ended.

Thus it was finally proved that the Luftwaffe could not sustain a damaging offensive against Britain with its conventional bomber force. Its inadequacy had become increasingly apparent as defences had improved and one important reason for this was Germany's lack of a heavy, powerful bomber aircraft available in meaningful numbers - the Luftwaffe still awaited mass production of a four-engined bomber to compare with those of Britain and America. It was choices made early in the Luftwaffe's existence which had brought this situation about.

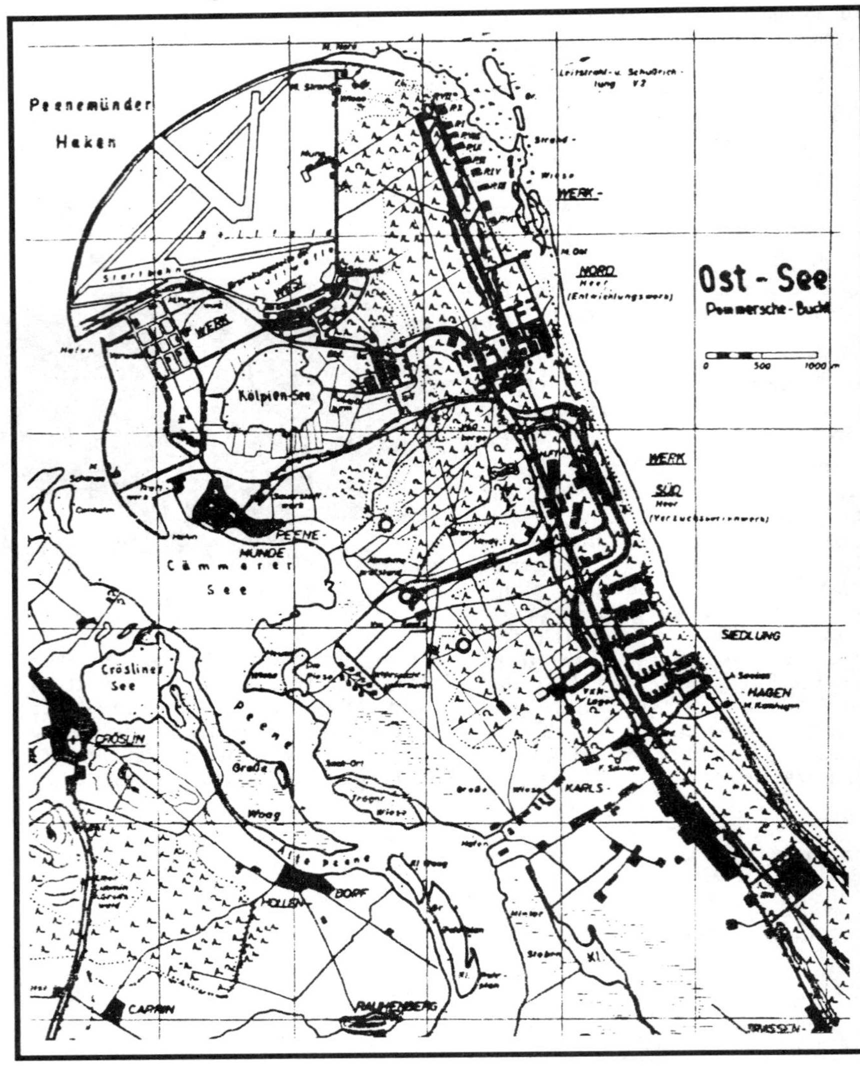

A German map of the experimental station at Peenemunde where the V-1 was developed. The airfield, Peenemunde West, is marked by its runways at the tip of the peninsula; the bombers carrying the flying bombs for air-launched tests took off from here. On the opposite side of the peninsula a series of ramps directed ground-launched test missiles eastwards out over the Baltic. Below the airfield, the fishing village from which the station derived its name nestles on the inlet of the River Peene

General Walther Wever, the first Chief of the Luftwaffe General Staff after its formal creation in March 1935, was a proponent of the heavy long-range strategic bombing aircraft, but he died in an accident on 3rd May 1936. At that time the German airforce lacked a satisfactory bombsight but some experimental successes had been achieved in accurate dive bombing experiments. General Wever's successors foresaw the main role of the Luftwaffe as that of a close support or tactical adjunct to the army. This and the need to build up a superior airforce quickly for pressing political reasons suggested manufacturing concentration on small and medium bombers, with a dive bomber capability so far as was practicable.

By 1939 this expansion had been largely achieved. The changing political scene then suggested the need for a heavy long-range bomber, but the dive bomber concept had taken such deep root that, incredibly, the specification drawn up for this heavy bomber provided for a dive bombing capability. This was in spite of the introduction of the improved BZA-2 conventional bombsight. The heavy bomber chosen for development was a Heinkel design, the 177, which, because of the stresses of the dive requirement, had a greatly strengthened airframe and the enormous flying weight of 32 tonnes. In order to reduce drag and to help achieve an adequate performance, the Heinkel engineers decided to make use of a revolutionary design in which four engines were combined in pairs, side-by-side fashion. Each pair was to drive one propeller, instead of employing the simpler design of four independently functioning propellers.

In practice, the use of the paired engines proved to be a source of constant trouble and even danger, since they repeatedly overheated and caught fire. As a result, crashes were common. Over three years of valuable time were absorbed in rectification of the technical problems arising and the only German purpose-built heavy bomber never went into large scale production.

The unconventional strategic V-weapons were seen by the German authorities as a means of delivering a long-range offensive against Britain on a decisive scale by methods other than the troublesome heavy bomber. This alternative means called for the application of an ingenious technology to a very specific need.

High priority had been given to both seemingly promising V-weapons' development programmes. Progress was, however, seriously impeded as a result of Allied bombing - particularly by a heavy raid on the Peenemunde experimental establishment carried out on 17th August 1943.

This research and experimental station was on the island of Usedom, just off the north east coast of Germany in the Baltic Sea. The German Army and the Luftwaffe jointly purchased its north-western peninsula in 1936 and set up upon it, near to the fishing village of Peenemunde, experimental firing ranges directed out over the Baltic. The research establishment grew up alongside these firing ranges.

Whilst the army experimented with the series of rockets which eventually resulted in the successful A4, subsequently christened V-2, the Luftwaffe, on their part of the establishment, independently developed the flying bomb. There was a good deal of interservice rivalry and both services feared that higher authority would close down their project in favour of the rival. Following the RAF bombing raid there was a progressive dispersal of manufacturing facilities for V-weapons, partially to rapidly-built underground factories at Nordhausen in the Hartz Mountains in Central Germany. However, the general disruption of the economy and constant changes of design were to have a crippling effect on the development and production of both weapons. The consequent hold-up to this introduction into operational service greatly reduced their ultimate military effectiveness.

Of the two, the V-1 weapon, each unit of which cost only about a tenth of that of a V-2, was used on a much larger scale and came into operation first. It crossed the coast of South East England on 13th June 1944, aimed at London, a week after the D-Day landings in Normandy, and six months later than programmed.

Its introduction was intended to be accompanied by a concerted effort. In a too optimistic directive of 16th May, Field Marshal Keitel, head of Hitler's OKW supreme headquarters, ordered that long-range bombardment of England was to commence in mid-June and that in addition to the ramp-launched flying bombs, air launching, long-range artillery and conventional bombers were to be simultaneously used. The start was, however, a fiasco; only ten V-1 missiles and thirty-three artillery shells were fired on the opening night. On 15th June the campaign finally got into its stride, the air-launched V-1 element making some contribution but the conventional bomber effort envisaged by Keitel's directive failed to materialise at all. In total, 6,725 of the flying bombs are known to have approached the coasts of Britain, just under 4,000 of these being destroyed by the defences.

Despite its vulnerability, the missile did have a serious effect both on public morale in the South East and in terms of the physical damage done. There was, too, a temporary, but marked, fall in the level of war production for over two months in the affected areas.

During late June and throughout July and August 1944, a V-1 bombardment of London was kept up by day and night. This forced a major civilian re-evacuation and disrupted life to an extent not experienced since the blitz raids of the second half of 1940 and first half of 1941.

The weapon caused much blast damage in built-up areas and the noise of its passage alone was calculated to cause alarm. It was impossible to tell at what point on its flight the V-1 would suddenly take an erratic dive to the destruction of itself and all in its vicinity. Despite its pre-set robot mechanism, there was a substantial number of shortfalls between the launch ramps and the target. Missiles damaged overland by the defences or caught by barrage balloon cables often did not conveniently explode in the air. Instead they came down on the heads of the defenders, or more likely their neighbours. The south-east corner of England between London and the coast opposite the launching sites was dubbed grimly "Bomb Alley" at this time.

By contrast, the V-2, of which 1,178 were successfully fired at Britain, was by its nature a weapon of much higher speed and one which penetrated deeper at its point of impact before exploding. It caused even more extensive local damage than the V-1, often blowing a deep crater thirty feet or more across. Its blast effect, however, tended to funnel upwards rather than spread in even concentric circles as was the case with the V-1.

It was not practicable for the defenders to shoot down such a high speed weapon, though it did sometimes explode in the air of its own accord owing to technical failure. Whilst its range was only some 200 miles, this embraced the whole of London from launching sites in Holland and from September the V-2 bombardment continued right up to the end of March 1945.

Allied armies broke out of their bridgehead in Normandy and at the beginning of September the last of the V-1 launching sites in France was evacuated by the V-1 ramp launch crews. From thence until the introduction of longer range flying bombs in March 1945, London was out of range of V-1 launching ramps.

As previously indicated, however, an alternative delivery method had been developed; that of launching from underneath bomber aeroplanes. This was first used on a small scale in June, but progressively expanded. It was developed determinedly into a larger scale operation once the launching ramps were evacuated.

Duncan Sandys (later Lord Sandys) was appointed Chairman of a "Flying Bomb Counter Measures Committee", specially set up when the quality of the V-Weapon threat became known. On the cessation of the cross-Channel bombardment after a perilous two-and-a-half months, the relief felt by those responsible for air defence and home security was almost boundless. Duncan Sandys and Herbert Morrison, then Home Secretary and Minister of Home Security, called a press conference at which Duncan Sandys used words which, in a short time, were to have a very hollow ring: *"Except, possibly, for a last few shots the battle of London is over..."* The very next day, 8th September, the first of the V-2 rockets to be launched at Britain exploded at Chiswick in London.

The existence of V-2 was not immediately revealed to the public. It was not until the German authorities, having first satisfied themselves of the success of the weapon in operational service, announced

The V-1 production line at the Nordhausen factory beneath the Hartz Mountains. The factory was captured by the US Third Armoured Division in April 1945

(Imperial War Museum. Copyright reserved)

its use in their radio broadcasts, that the British Government acknowledged that it had arrived. The announcement, on 10th November, said that the United Kingdom was under rocket bombardment, but did not refer to London. All the while, in the South East, the mysterious explosions continued to occur and were obviously of serious magnitude and due to enemy action.

In its September 1944 issue, the Civil Defence Journal "ARP News" took a more serious view than Duncan Sandys of the potential of the air-launched V-1. It wrote that "Robot bombs may not be confined to Southern England" and advised on the preparations which it thought ought to be made to meet the menace in other areas. This was ominous and prophetic since the known range of the V-1 was only a little over 150 miles from its launching point.

Air Marshal Roderic Hill (later Sir Roderic Hill), then Chief of the Air Defence of Great Britain, described how, as early as 8th July, radar operators covering the approaches to London were observing flying bombs approaching from a new direction - the east! After a few weeks it was established that these missiles, which came only at night, were being launched from specially equipped Heinkel 111 aircraft. This method of launching, evolved from experimentation carried out at Peenemunde during the winter of 1943-4, added an extra perspective, giving the flying bomb a new flexibility and threatening a large part of the country - including the North of England.

Rescue workers search the debris for victims after a V-1 attack on southern England (Imperial War Museum. Copyright reserved)

Since the battles of Stalingrad and Alamein in the closing months of 1942, the war had gone badly for Germany, never more so than in the autumn of 1944. On the Eastern Front, Russian troops crossed the River Danube south of Budapest on 29th November. In the West, Allied troops were poised on the borders of Germany itself. They were temporarily held up by the need to open the port of Antwerp and through it to accumulate the necessary supplies and reinforcements for the final push to end the war.

In the early hours of 16th December, a total of 20 German divisions under Field Marshal Von Rundstedt struck, in a major offensive launched in foggy weather out of the Ardennes forest against the 8 Corps of the US First Army. Allied air support, which otherwise could have blunted the attack, was impossible in the prevailing weather and the weight of the German assault caused a confused American withdrawal. The German advance was not finally halted until Christmas Day. The enemy had advanced 40 miles, almost reached the River Meuse and had besieged an American garrison in Bastogne. Von Rundstedt failed by a matter of 70 miles to gain the objective which Hitler had set for him - the now-vital Allied supply port of Antwerp.

When the Ardennes offensive was approaching its climax on 24th December, the air-launched V-1 was finally turned against Northern England. Dr Goebbels' Propaganda Ministry, ever quick to make the most of a military success, linked the two events. German broadcasts, including those in their English language service, referred to the "extending of our long-range weapon bombardment to Manchester in conjunction with our West front attack." Two temporary successes were made to sound like a reversal of fortune.

Chapter 3

A NOISY LITTLE AIRCRAFT

From mid-1942 onwards evidence began to reach the British Military Intelligence Service of German experimentation with a long - range robot - controlled missile. What was not at all clear was that more than one type of missile was involved and this confusion clouded a clear understanding of the information gleaned for months to come. Duncan Sandys, himself an ex-artillery officer, was appointed to take charge of a special investigation group. This was constituted on 20th April 1943, to monitor these enemy developments on behalf of the War Cabinet. The Air Ministry was asked to direct RAF photographic reconnaissance units to keep watch on likely long-range weapon development sites, and as early as 9th May Duncan Sandys visited the RAF Photographic Interpretation Unit at Medmenham, near Henley-on-Thames. It was even then envisaged that the enemy secret weapon might take the form of a pilotless aircraft. He therefore made a point of talking to the chief photographic interpreter of the aircraft interpretation section, Flight Officer Constance Babington-Smith. She had seen nothing in her searches up to that time to provide any identification of such a weapon. Important evidence was, however, spotted on photographic covers taken over the site at Peenemunde in late June. Two large rockets were seen lying on their sides in the open. This was considered sufficient to justify the mounting of a major RAF bomber raid on the Peenemunde research establishment and the night of 17th August was chosen for this.

During the course of the raid, many foreign workers there were killed or injured. Amongst them were Polish manual workers who had provided valuable information to their underground army in Poland, which, in turn, passed it on to London. Some of this information concerned a noisy little aircraft that had been seen flying out to sea from the Peenemunde ranges.

On 27th August a message from an agent or "source" high in the weapons department of the German War Office gave the first definite confirmation that two distinct weapons were being tried out; a pilotless aircraft which was identified as Fi103 and a rocket projectile identified as A4.

It was not until 28th November that photographic confirmation of the pilotless aircraft finally became available. Constance Babington-Smith, re-examining previous aerial photographs of Peenemunde, located a tiny object in the form of a cross, in an unusual position, but which could well be a robot aircraft. On the very same day, favourable weather conditions enabled further aerial photographs of the establishment to be taken. Examination of them revealed a similar small shape at the base of what was probably a launching ramp pointing out to sea.

In the late spring of 1942, General Erhard Milch, the German Secretary of State for Air and Inspector General of the Airforce, had been invited to the Army Research Establishment at Peenemunde to attend an early firing demonstration of the A4 rocket projectile. He was so envious of the progress made by the army, who claimed that their rockets would have range enough to bombard London, that he issued a Luftwaffe requirement for an expendable automatically piloted aircraft that could reach the city from German-occupied France, and thereby match the army's weapon. The concept of a pilotless aircraft to be used as a missile was by no means new and had been experimented with as early as the First World War. What was new was the revolutionary technology which was about to be brought together for the first time to meet this specific and urgent need.

On 10th June a meeting of a Scientific Committee of the German Air Ministry was held, chaired by Milch. Dr Fritz Gosslau of Argus Motorengesellschaft of Berlin was present to describe the success of experimental work previously carried out by himself and Dipl-Ing Paul Schmidt on the practicability of a small, cheap, pulse ram jet engine. He suggested that this could become the motive power for an effective flying bomb. This accorded exactly with the Luftwaffe requirement for propulsion of the pilotless aircraft type of missile envisaged. Milch showed such enthusiasm for the idea that orders were quickly placed for the undertaking of design and experimental work.

Gerhardt Fiesler, a First World War pilot who had become an aircraft manufacturer of high repute, drew up a general specification and an outline scheme to meet the overall requirement. This was accepted by the Luftwaffe High Command and a team of engineers under Dipl-Ing Lusser commenced work on the detailed design of the fuselage and control system. Professor Willi Messerschmitt was appointed a consultant aerodynamicist.

The Fiesler aircraft manufacturing company was made responsible for the project as main contractor. In accordance with German aircraft industry practice, the machine was given the title Fiesler Fi103. To camouflage its purpose for security reasons, it was to be called Fern Ziel Gerat (Long Range Target Apparatus) FZG76. Development of one of the key components, the cheap, mass-produced automatic pilot, was placed in the hands of the Askania Werke AG of Berlin. This company was amongst the foremost German precision instrument manufacturers and had, in 1933, been licensed by

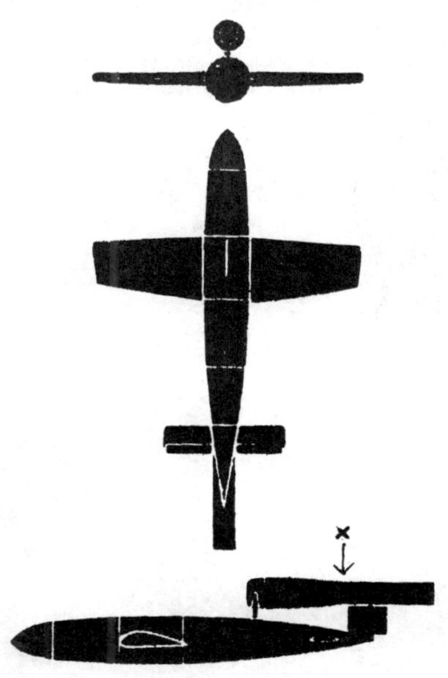

An official drawing of the German pilotless aircraft representing a wing span of 16ft., length 25ft. 4½in. Probable bomb load is 1,000lb.

This remarkably accurate representation of the V-1 appeared in British newspapers as early as June 17th 1944, soon after the start of the ramp launching operations. It was incorrect in that the wings were shown to have a tapered chord. For simplicity of construction, all flying bombs had a parallel wing chord

Sperry Gyroscope Inc. of Brooklyn, USA, to manufacture and sell their automatic pilot system. Sperry had furnished technical data, tool and production drawings and details of manufacturing methods to get Askania started in the gyroscopic automatic pilot business.

By May 1943, powered launching trials were being carried out from ramps on the Luftwaffe side of the Peenemunde establishment and the missiles launched were tracked out over the Baltic by German radar. As successive tests proceeded, accuracy and reliability improved to an acceptable level. Nevertheless, a relatively high rate of launch failures and shortfalls of operational missiles was accepted as inevitable. The whole assembly was designed for simplicity and mass production by relatively unskilled labour. Its single operational flight was also its test flight. The need to avoid the use of scarce raw materials imposed another constraint on the designers.

The V-1's Argus pulse duct motor was well suited to its application, but it was enormously wasteful of energy. For every mile of flight it used up a gallon of petrol, about as much as that used by an RAF Lancaster bomber whose piston engines developed nearly 6,000 hp between them. Its efficiency tended to fall progressively as the flight of the missile proceeded. The air inlet grid and flap valves of the pulse duct had a designed working life of only 30 minutes, for, despite a protective baffle, flame from the combustion chamber usually caused them some burning and erosion in operation. Allied fighter pilots approaching the missiles in flight even reported the escape of flaming gases from splits which had been burned through the shell of the ducts. Such escapes of energy not only further reduced the efficiency of the motor, they also tended to upset the automatic navigation of the bomb. In normal circumstances, the small loss of pulse jet efficiency during flight was offset by the lightening of the missile as its petrol and compressed air stocks were used up. Approximately 50 litres of air, as measured at atmospheric pressure, were used up from the compressed air supply for every mile flown.

One consequence of the variation in the quality of the mass-produced Argus duct motors was a wide variation in the speed of the individual missiles. Speeds between 350 and 400 mph were reported at various times. An average of around 360 mph was normal within the wide range. A minimum launching speed of not less than 150 mph was necessary in order to enable the pulse duct thrust to take the bomb away without its stalling. A motor with a better thrust or one which had been running longer would stand a better chance of getting the bomb away to a successful air launch. Once the missile picked up flying speed, this rose to a peak before falling gradually as wear and tear reduced the efficiency of the jet. The uncanny resemblance of the sound of the pulse jet to that of an unsilenced two-stroke motorcycle engine frequently excited the comment of observers. The designed frequency of the pulse explosions was determined by the length of the duct pipe. This was fixed to give 45 pulses per second, 2,700 per minute - a similar rate to that of some motorcycle engines.

This low-flying V-1 has just been launched from a ramp. The capture of the launching ramps by the advancing Allied armies was the prime reason for the rapid expansion of the German air-launching programme (Imperial War Museum. Copyright reserved)

Externally, the air-launched V-1 was sprayed in a dark olive green paint on its upper surface and a light blue (hellblau) on the lower. The front motor support strut fairing, tailfin, rudder and the underside of the jet duct pipe were also light blue. A ragged line of paint interface ran along each side of the fuselage and jet pipe. Surfaces were often overdaubed with grey to improve camouflage. Operational and handling instructions were stencilled on, mostly in black lettering on blue parts of the surface of the fuselage. Letters 15mm or 20mm in height were used, but no national markings of any sort were ever carried, implying that the Germans regarded the machine as a missile rather than as a pilotless aircraft. A protective coat of red or grey primer was sprayed on to the internal components as they passed along the production lines. Sample fragments from the missile which fell at Turton, Lancashire, show signs of the red primer. The wing components from the Newport, Shropshire, bomb, now held at the RAF Cosford Museum, show traces of the grey primer. Since major V-1 components were only finally assembled at flight preparation, the painted surfaces did not always match. There were instances reported of flying bombs entirely without paint. As early as 25th June, principal Civil Defence officers in the South East were circularised with an instruction to notify higher authority of any instances of these "paintless flys". Whether these resulted from reasons of economy or in order to bypass a production bottleneck was never apparent, nor is it certain whether any of the unpainted bombs reached air-launch units.

One common misconception regarding the V-1 is that the motor always cuts out at the commencement of the dive. There was a general faith amongst the public that "It's all right so long as the noise keeps going," a faith that was sometimes sadly misplaced. Very often the motor did cut out as the dive started, owing to the effect of the remaining fuel being thrown away from the petrol tank outlet pipe. If, however, the dive were a very shallow one, this might not happen and the motor could continue to run. On other occasions, such as that observed off Reads Island in the Humber, the bomb could spiral down through mechanical defect with its motor running. Sometimes the motor would cough on a mixture of petrol and air from the fuel pipeline before cutting out. A distinct whine was sometimes heard during the descent of the missile as its remaining compressed air escaped from the spray nozzles and the severed control pipes. The fears expressed by the Regional Civil Defence Headquarters at Preston that the finding of one of the compressed air bottles of the Worsley bomb might have constituted a danger to the rescue teams were ill-founded. Even if the spherical bottle had not been crushed by the explosion of the warhead, any residual air it contained would be down to atmospheric pressure.

Through their ingenuity, the German designers of V-1 had neatly combined the features of both a bomber aircraft and long-range artillery into a reasonably accurate, cheaply produced and extremely destructive stand-off weapon. It was intended primarily for use against England and made possible offensive action on a decisive scale once more. In the event, it was introduced just too late for it to be fully effective. For both technical and military reasons, events had started to overtake the flying bomb by June 1944.

Chapter 4

'PIGGY BACK'

Towards the close of 1943, pressure groups within the German High Command began to busy themselves with the question of control over the intended V Weapon campaign against England, to which both Army and Luftwaffe were to contribute. Field Marshal Von Rundstedt, the exceedingly able and respected Commander-in-Chief - West, persuaded Hitler to place control in his office, effectively giving it to the Army. The order placing this responsibility was contained in a Fuhrer Directive dated 1st December 1943. This went on to order the activation of a headquarters of Army Corps status - 65 Corps - under Lieut. General Erich Heinemann, an officer experienced in artillery, but then 67 years old. His Chief of Staff, on whom much of the executive responsibility devolved, was to be another army officer, Colonel Eugen Walther, who, as it turned out, lacked the experience and imagination necessary to exploit the full potential of the air-launched V-1.

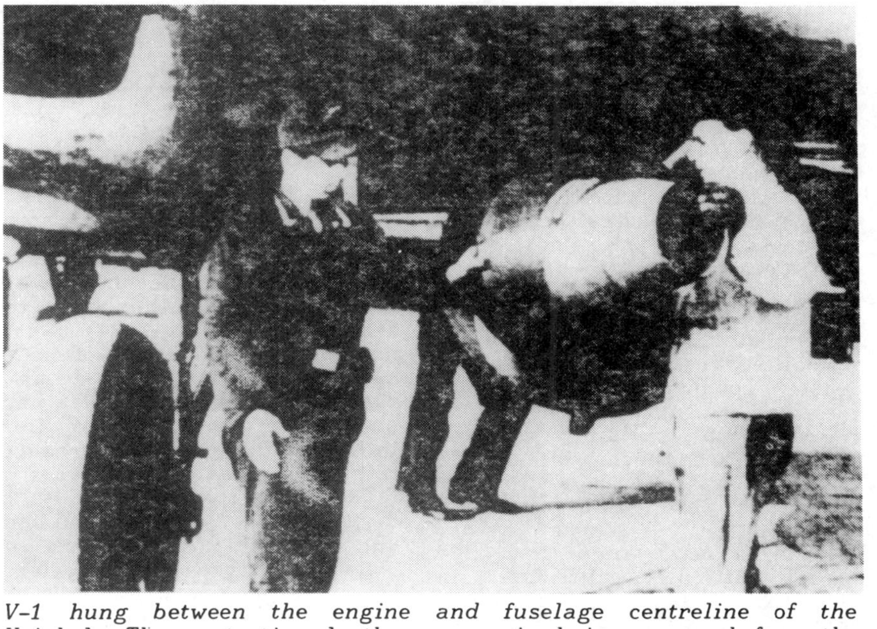

V-1 hung between the engine and fuselage centreline of the Heinkel. The protective leather cover is being removed from the airlog windmill propeller in readiness for takeoff

The headquarters, staffed by Army and Luftwaffe jointly, was established on the outskirts of Paris and came into existence on 15th December 1943. It was to control all V Weapons used against England; V-1, V-2 and an intended ultra-long-range fixed multi-barrelled gun, V-3.

A Luftwaffe Regiment under the command of Colonel Max Watchel was created to operate the V-1

A Luftwaffe airfield tractor tows a V-1 on its trolley from the camouflaged armourers' hangar towards a waiting Heinkel. The wings of the missile, which have already been fitted, will have to be removed again before loading on to the aeroplane. The wings were easily detachable by removing single bolts which held them on to the tubular mainspar

launching sites and went into training at Zempin on the Baltic coast. To disguise its purpose, this was designated Flak (Anti Aircraft) Regiment 155(W); the W standing for Watchel.

Meanwhile, construction started on intended launching sites in Northern France aligned on London, Southampton and Bristol. These were cumbersome concrete structures impossible to hide from the air. The number of foreign labourers and French contractors employed in the building work ensured a flow of information concerning them back to British Intelligence. Allied bombs soon severely disrupted the work. With hind-sight, it seems a pity that the Allied air attacks on these initial V-1 launching sites, christened by the British the "Ski" sites, on account of their shape, were so hastily carried out. As sites were destroyed, the Germans had to find much more easily concealed and less complex methods of launching. Very quickly the so-called "Modified" ramps were designed, easily constructed out of steel sections set up on trestles. The amount of building at the "Modified" launch sites was kept to a bare minimum and the sites were easily concealed in militarily secure areas. The Allied airforces had great difficulty, therefore, in locating them. It is worth noting how well this lesson was learned by the Germans, in that when V-2 was eventually used, its launching was achieved without the establishment of any permanent sites at all.

Because of the limited range of V-1, the procedure for continuing the attacks in the event of loss of the launching sites on the French coast was carefully considered by the German planners. In order to maintain a large-scale attack, the range of the weapon itself would have to be extended. Alternatively, the bombs could be launched from aircraft. By using this method, not only could the V-1 offensive against England, codenamed "Rumpulkammer", be kept going, but a greater measure of flexibility could also be introduced into the operation. Attacks from launcher aircraft could be directed so as to outflank and confuse the defenders. Of potentially greater importance still, targets well outside the range of even the longest-range flying bomb envisaged would be open to attack.

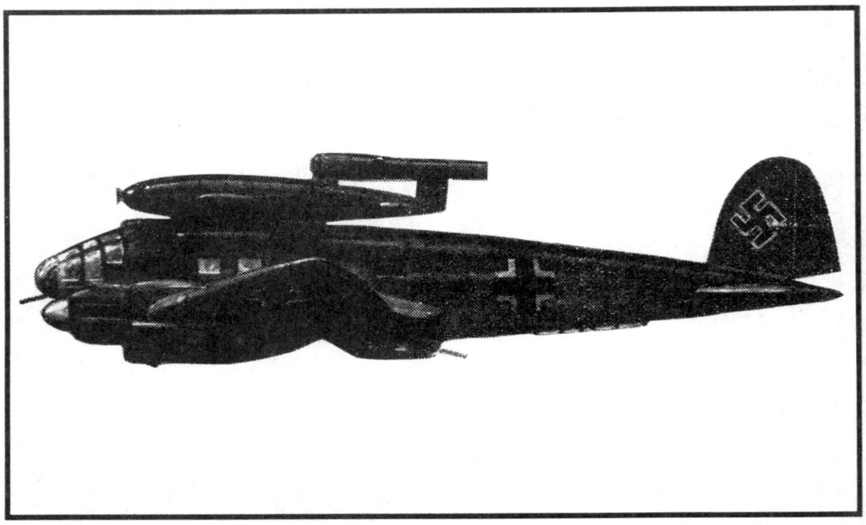

The imaginative way in which a press artist portrayed the air-launched V-1 in September 1944. The idea was taken from the Mercury-Mayo flying boat combination used for long-range mail flights in pre-war days. Only later did the British public learn that the flying bomb was carried under the starboard wing of the Heinkel

Air launching trials carried out at Peenemunde had not only proved the feasibility of launching flying bombs from aircraft, they had also established that the best German aeroplane then available for the purpose was the tried and reliable Heinkel 111.

On the face of it, the V-1 appeared to be an unwieldy load for any aeroplane to carry and to launch from the air. The method adopted was the only one then really feasible. The flying bomb was suspended from the wide wing root as close in as possible to the centre line of the aeroplane. It was to be expected that the handling characteristics of the loaded Heinkel, particularly during take-off, would be made difficult, even hazardous. Surprisingly, however, pilots reported that stability was not unduly affected when carrying a V-1 and that take-off did not present any problems, being accomplished in 800-900 metres (900-1,000 yards). The drag caused by the missile, once airborne, was corrected by trimming the rudder about 1°. The speed of the Heinkel at a height of between 100 and 200 metres (350-700 feet) with a V-1 underneath was reduced by only about 20 Kph (12½ mph) compared with the unladen speed.

Heinkel III H-16, Works No.161600, of 2/KG53 on Ahlhorn Airfield. The machine is readily distinguishable from the H-20 types used by its elongated upper gun position canopy. The mottling on the V-1 can also be seen

It was not difficult for British Intelligence quickly to identify the type of aircraft and the unit involved in the air launching. Initially, however, they incorrectly believed that the V-1 was in some way carried above the fuselage of the bomber. An early clue was apparently available in aerial photographs taken of Venlo airfield which showed a Heinkel 111 with what appeared as a long, unidentified object above the fuselage, extending from the nose back to a point in the rear of the trailing edge of the wings. It was thought that this might be the covering on a launching device. A speculative drawing was prepared, indicating launching rails. This speculation filtered down to the British press who, in turn, prepared their own artists' impressions of how they thought the air-launch operations were carried out. Typically, the London Illustrated News showed an artist's impression of the launching of a V-1 carried piggy-back fashion on rails above the fuselage of a Heinkel. The suggested method disregarded the effects of the jet tailflame of the missile at launching on the fin and rudder of the host aeroplane. This imaginative concept probably owed its origin to the Mercury-Mayo Mailplane composite flown pre-war, which employed a smaller floatplane carried above a flyingboat. The latter launched the former for long-range mail flights whilst airborne. It is worth noting here that an elaborate scheme was planned later for launching the V-1 from the back of an Arado 234C jet bomber. This involved a special cradle on the back of the parent aircraft in which the V-1 sat until the launching zone was reached. A series of hydraulically-operated arms then raised the missile clear of the top of the aircraft for launching. This scheme was still at the planning stage when the war ended.

The concept had its humorous side. A cartoon in the Aeroplane Journal of November 1944 showed a flying bomb with its engine running, jammed on rails above the cringing pilot of a juddering Heinkel. The pilot called to the unseen bomb-aimer that he did not care how the missile was got unstuck, but at all events it had better be QUICK!

To the aircrew of the Heinkels engaged in the air-launch operations this would seem wry humour indeed. Nevertheless, given the necessarily hazardous nature of the operations in which they were engaged, they had got an aircraft which, though obsolete, was comparatively well suited to the task.

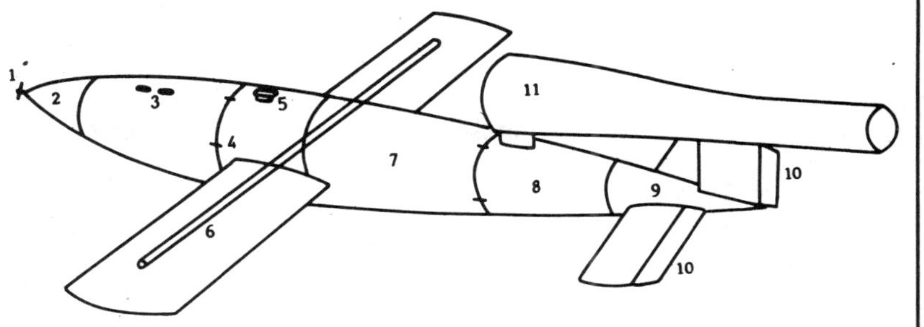

The air-launched V-1

(1) Propeller for airlog. This windmilled in the slipstream of the forward flight and tripped an airlog tachometer. After about 56 km (38 miles) of flight, the airlog armed the electrical fuses of the warhead.

(2) Alloy nosecone containing directional compass.

(3) Warhead section with fuse pockets. Explosive fillings varied from the 52A, a standard type used in some general purpose HE bombs, to the Trialen 105 filling, a special high-blast explosive mixture. Each air-launched missile carried 850 Kg (1,874 lbs) of explosive, which composed 94% of the total weight of the warhead.

(4) Fuel tank, with a capacity of 500 Kg (150 gal or 682 litres) of B-Stoff, the standard German road vehicle petrol.

(5) Suspension lug for attachment to Heinkel.

(6) Tubular mainspar attaching wings.

(7) Compressed air storage. The air which forced fuel out of the petrol tank was stored in two mild steel spheres, bound for extra strength with layers of steel wire.

(8) Central compartment containing the automatic pilot (Steuergerat), the petrol supply controller, altitude controller, airlog and the actuating controls for the rudder and elevators. Since there were no ailerons, all control in flight was exercised through adjustment of the rudder and elevators. If one wing dropped, the rudder steered away from that wing, producing greater velocity over it and thus restoring it, through greater lift, to a horizontal plane.

Two detonators set between the tailplanes were fired at the termination of the flight on a signal from the airlog. These released hinged flaps, called spoilers, under the tailplanes. Their effect was to increase the lift of the tail and put the missile into a diving posture.

(9) Control rams.

(10) Tailplane control surfaces.

(11) Argus impulse duct, ram jet motor. When first started up, the pulse jet was fired by a conventional bosch sparking plug on top of the combustion chamber. The plug took its power through an umbilical cable from a starting magneto in the carrying aircraft.

The Heinkel 111s were converted into flying bomb launchers at Oschatz. Modifications involved removing the internal bomb racks and the ventral gun position. The starboard wing root was strengthened with metal sheeting around the inboard fuel tank. A shallow, streamlined carrier pod was fitted under the starboard wing root, between the fuselage and the wheel housing. This had two lugs to engage the T-piece at the top of the V-1. The 2,180 Kg (4,796 lbs) weight of the missile was distributed by this carrier across the inboard wing spars. A 21 cm diameter windmill vane was attached to a tube projecting from the leading edge of the starboard wing, outboard of the engine. Its rotations were counted on an airlog tachometer which formed part of a release mechanism called Zahlwerk. This instrument was contained in a box placed slightly above and in front of the pilot. A duplicate box was fitted at the bomb-aimer's position on the starboard side of the fuselage, with the addition of a black button known as Anstellknopf and a red button known as Abstellknopf. The former enabled the bomb-aimer to start the motor of the V-1 and the latter to stop it in case of pre-ignition or other failure. A short tube out of the starboard side of the fuselage directed a looped cable to the pulse jet spark plug of the missile. This carried the current from a magneto in the aeroplane via the Zahlwerk to start the pulse jet. Beside the bomb-aimer's Zahlwerk box a lever released the missile. An emergency release lever was situated above the pilot's head.

Each Zahlwerk box was only 20cm high and 12cm wide. In the centre was an illuminated window which displayed a set of digits.

The fuel-carrying capacity of the aircraft was 748 gallons (3,460 litres), which gave an endurance of $5\frac{1}{2}$ hours, adequate to cover the maximum of 4 hours required on flying bomb launching operations. The overall weight of the unladen Heinkel was materially reduced by the air-launch modifications. In all operational conversions, the flying bomb was carried underneath the starboard and not the port wing.

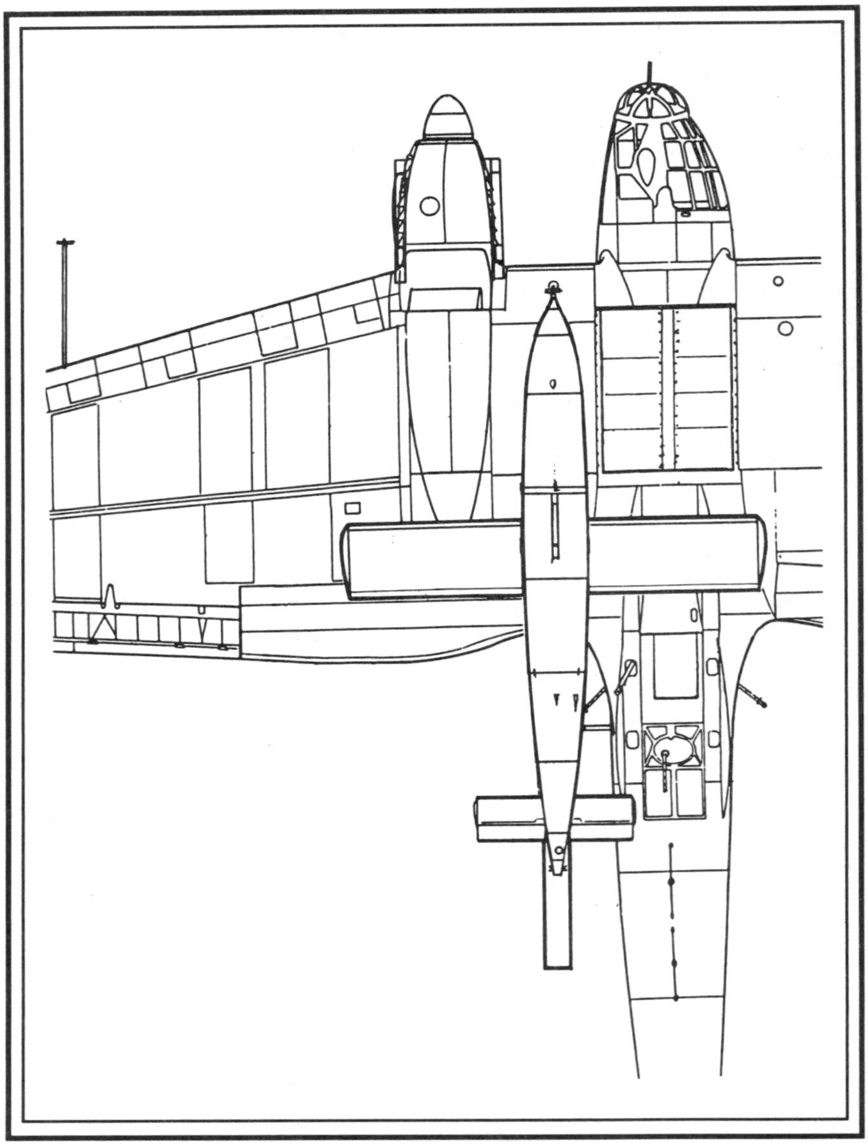

A V-1 beneath its carrier Heinkel (Drawing by Frank Leyland)

When fitted to its host aircraft the port wing of the V-1 cleared the underside of the fuselage by only 8-10cm. Lateral stability was maintained by two vertical metal struts, having on their lower ends horizontal spring-loaded wooden buffer blocks which rested on the upper surfaces of the missile wings.

Conversion training of aircrew assigned to flying bomb launching operations was initially carried out at Karlshagen airfield, Peenemunde. This consisted of instruction in flying with and releasing the V-1 in a ten-day course conducted by Erprobungskommando Karlshagen. Three flights, each of one hour's duration, were made by the conversion trainee pilots. The first of these was to give the pilot experience of the aircraft. Only the pilot took part and no flying bomb was carried. The bomb-aimer, meanwhile, received about half an hour's instruction on the ground in the operation of the release gear. In the second flight, a flying bomb without a live warhead was carried and the whole trainee crew flew, accompanied by an officer pilot instructor. During this flight the starting procedure, aiming and release of the missile were practised. On the final training flight the crew practised a launch on their own without an instructor present. Several crews were under instruction at the training school at the same time.

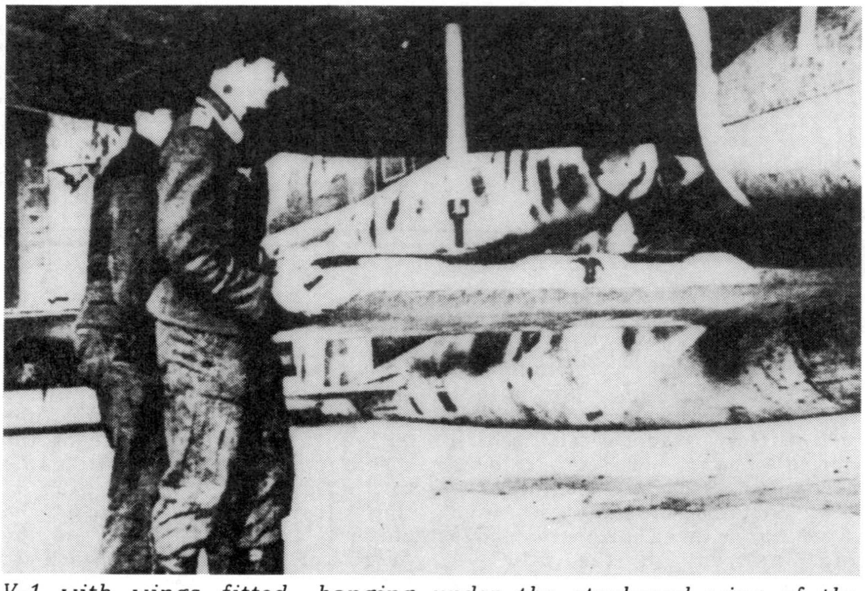

V-1 with wings fitted, hanging under the starboard wing of the Heinkel. The spring-loaded steadying strut from the aeroplane rests its buffer block on the upper surface of the wing of the V-1.

Some, at least, of the aircrew members underwent a ground-launch flying bomb course. Hans Hoehler qualified as an aircrew radio operator in September 1944. On 2nd October he was posted to KG53 and sent on such a course, arriving with 29 other airmen at Zempin on 11th October. After a week's delay, their course started with detailed instruction on the technicalities of the V-1. To end the course, each syndicate of five trainees set up and fired a missile from the Zempin launching ramps, assisted by the crews of Flak Regiment 155

(W). Hoehler's syndicate achieved the most accurate shot, within 1 Km of the target at near maximum range.

In Hans Hoehler's case, the ground-launch course was given as an alternative to the air-launch course undertaken by most of the aircrew. It included instruction in air-launch aspects of V-1, but does illustrate the inadequate and stop-gap nature of the training given to air-crew during the October/November period of rapid expansion into KG53. He was finally posted back to KG53 head-quarters at Bad Zwischenahn on 16th November and from there joined 3 Staffel at Vechta.

The first crews trained, during May and the first half of June, 1944, were formed into III Gruppe Kampfgeschwader 3, which Geschwader carried the honorary title "Blitz". It had only III Gruppe, consisting of 7, 8 and 9 Staffeln. This proceeded to French airfields and, despite some losses on the ground in a fortuitous Allied bombing attack on Beauvais Tille, commenced air-launch operations on a limited scale, almost in the nature of operational training. Nevertheless, launchings took place over a wide stretch of the Channel coast between Fecamp and Flushing. The earliest aiming point was London, but later both Southampton and Gloucester were included before III KG3 was withdrawn to Gilze Rijen in Holland and Venlo on the Dutch-German border towards the end of July. Hauptmann (Captain) Martin Vetter commanded the Gruppe. He had previously been in command of II KG51.

The first air-launched missiles which the British were able to identify appeared on 8th July. Coastal radar picked up the tracks of incoming flying bombs from the direction of the Belgian coast where no launching sites were known to exist. These and following missiles, which were launched only at night, came in up the Thames estuary and ominously outflanked the existing flying bomb defences.

Whilst III KG3 formed part of the Luftflotte 3 of the German Air Force for administrative and supply purposes, it came under the operational command of 65 Army Corps which embraced all flying bomb operations. On 26th June, the Corps had authorised the firing of 12 ramp-launched flying bombs at Southampton. Certain German agents in London, who had been ordered to report the fall of flying bombs, had already been turned into double agents by the British. Under their orders these agents signalled that substantial damage had been done in Southampton and even that a fighter squadron stationed there had been forced to transfer to another airfield. This bluff worked and 65 Corps decided to redirect a part of the air-launch effort to Southampton. The purpose of the bluff was to divert flying bombs from the vast built-up area of London to a relatively small and difficult to hit target. It had to be one sufficiently attractive to the Germans and Southampton, as a major supply port for the Allied armies in France, met this requirement.

With the authority of the Commander-in-Chief - West, now Field Marshal Von Kluge, III KG3 were ordered to attack Southampton, whilst maintaining the offensive against London. The intermittent attack lasted until the end of July, and at least ninety bombs were directed towards the south coast port by Heinkels. As was anticipated by the British authorities, the fall of the missiles was widely dispersed, though some damage and casualties did occur in both Southampton and Portsmouth. Some reinforcement of the anti-aircraft defences, particularly on the Isle of Wight, was made necessary and this for the British was an unwelcome result. Winston Churchill's scientific adviser, Professor F A Lindemann (later Lord Cherwell), aware of the success of the deception, urged Herbert Morrison, as Minister for Home Security, to "commiserate with a South Coast Town, indicating that the (air-launched) attack had been a success". Morrison replied that "politically it would be dangerous in the extreme" and "would be known to be untrue so that doubts would be cast upon the accuracy of the government's statement generally".

FLY-BOMB LAIRS IN BELGIUM?

From VICTOR LEWIS,
Daily Sketch' Air Correspondent

GERMANY is believed to be using a small group of new flying-bomb launching sites to continue the attacks on London and Southern England.

They are probably located in Belgium, but some could even be as far north-east as Holland if they were placed at the extreme limit of what is believed to be the maximum range of the flying bombs.

The severe pounding to which the Pas de Calais sites have been subjected by Allied bombers may have necessitated this switch.

It is known that several of the sites in France have been wrecked and a great many more so damaged that they will be out of action for some time

Lancasters Hit Ramps

Lancasters of Bomber Command yesterday afternoon resumed the attack on flying bomb installations in Northern France and returned without loss.

Flying bombs crossing the South coast yesterday by various routes, were chased by fighters which made many "kills." In one district the robots were reported at more frequent intervals than usual.

A train carrying flying bombs was derailed when the Vaucour railway bridge on the Metz-Verdun line was blown up by French patriots, say authoritative French sources.

German troops were immediately drafted to the spot to prevent anyone from approaching the wrecked train.

This press report of 13th July 1944 of V-1s coming from the east was the first reference to an air-launched attack, although this is not stated in the article. Meanwhile, the ramp-launched campaign from Northern France grew in intensity

A Heinkel of 9/KG3 (later 1/KG53) with bomb suspended, ready for takeoff

In addition to those aimed at London and Southampton, KG3 evidently launched 20 missiles at Gloucester, but these were so widely scattered that the British authorities had no idea at the time that this city had been an aiming point.

Two contemporary maps drawn up by the British Intelligence Service show the fall of those known air-launched missiles which crossed the coast aimed at London and escaped destruction by the defences. The first shows those launched between early July and 5th September, the mean impact point of which is north-east of the city, with a spread over a radius of about 25 miles. These missiles were launched by aircraft operating mainly from Gilze-Rijen and Venlo, which airfields were evacuated on the approach of Allied ground forces on 5th September.

The second map shows similar flying bombs aimed between the recommencement of launching, on 16th September, and the final termination of air-launch operations on 14th January 1945. In this case, the spread of the impacts is much wider, extending to a fifty mile radius of London.

On the evacuation of Gilze-Rijen and Venlo, airfields were available to III KG3 in north-west Germany, and these were chosen in preference to others further to the north-east of Holland because of the increasingly difficult transport and supply situation. Further, it was considered necessary to spread operations over a wide range of airfields to minimise the effect of Allied bombing and night intruder operations.

In October, a major reorganisation took place. A part of Kampfgeschwader 53, a unit which had spent two years on the Eastern Front, had been brought back to Germany in September and put through an air-launch course at Karlshagen, Peenemunde. This was then posted to the airfields in north west Germany to amalgamate with III KG3. The combined unit was given the designation I Gruppe Kampfgeschwader 53 "Legion Condor", and remained under the command of Martin Vetter. Subsequently, the remainder of KG53 was brought back from the Russian Front and put through air-launching instruction at Karlshagen and at Grottau. Additional bomber personnel for reinforcement were plentiful as by now the conventional bomber arm in the west was in severe decline. Similarly, Heinkels were readily available for conversion into flying bomb carriers as units such as KG27 were transferred to other roles and re-equipped, or were disbanded altogether. These reinforcements, when posted into the new KG53, were divided into numbers 4 to 9 Staffeln to make up a II and a III Gruppe. Together they brought the Kampfgeschwader up to full nominal strength of about a hundred aircraft, some sixty or seventy of which could be expected to be available for operations on any night.

II KG53 became operational about the middle of November. It ran into serious trouble in mid-December when twelve aircraft were lost due to accidents in only two operations. For some reason a number of V-1 bombs detonated under the aircraft just after take-off. Operations of the Gruppe were suspended for a fortnight while an investigation took place.

At the end of November, III KG53, commanded by Major Almendinger, moved into airfields close to the Danish border to commence operations. An alternative landing ground was set up at Husum to the west of Schleswig, for returning aircraft to use in the event of interference from Allied intruder activity.

P-planes from Belgium

Halifaxes and Lancasters of R.A.F. Bomber Command, with fighter cover, yesterday attacked long-range weapon supply depots in Northern France

IT is believed that the enemy was using a new northerly flying-bomb base yesterday.

Most of the bombs which were launched in a series of short attacks appeared to come over the French coast between Calais and Dunkirk

Observers think that most of them were discharged from sites in Belgium.

If the enemy attempts to continue the bombardment from Belgium (writes an Air Correspondent) it will mean that the bombs will have a greater sea distance to cover, and our aircraft patrol will have more time and space in which to attack them.

Free from A.A. fire

Our aircraft would also have an opportunity of tackling the bombs free from enemy A.A. fire.

South-Eastern and Eastern England are within range of flying-bomb attacks from the Belgian coast in the Ostend area, but clearing the Pas de Calais will rob the enemy of numerous sites that were prepared for constant and intensive attack on this country.

He is not likely to be so well prepared in Belgium and will not be encouraged to make expensive installations there in view of the threat of the advancing Allied armies.

From other sites yesterday only a score or so of bombs came over in ten hours. Formerly a similar number usually came over in one hour.

Even as late as 1st September 1944, the press continued the myth of V-1 launching sites in Belgium. On this day, ramp-launched attacks against Britain ceased until March 1945, when a few long-range flying bombs were launched from ramps in Holland, directed at London. Meanwhile the air-launched offensive built up.

Crew members of No.1 Staffel of No.1 Gruppe of Kampfgeschwader 53 walk past one of their Heinkels on Varelbusch Airfield, from which they operated against England. The V-1 in its position under the starboard wing of the parent suggests that the crews are preparing for a very early start on their night raid and that darkness must fall soon. Each member wears a life jacket around his chest and shoulders

The deployment of the complete Kampfgeschwader 53 placed its headquarters at Bad Zwischenahn and separated the Staffeln to:

AIRFIELD	STAFFEL OF KG53	GRUPPE
Varelbusch	1	I
Ahlhorn	2	I
Vechta	3	I
Bad Zwischenahn	4	II
Jever	5	II
Wittmund	6	II
Leck	7	III
Schleswig	8	III
Eggebek	9	III

but airfields at Handorf and Hespepe were also used.

In the earlier operations of KG53, navigation to the launching zone was by dead reckoning with the help of visual beacons. The flight path from the airfields to the coastal departure point was marked by a series of low-power, flashing morse signal lamps. These were placed on the ground to mark out the route before each raid. On the coast, pre-arranged fixed searchlights and flak shell-bursts were sent up for the guidance of the navigators. This method worked well at the time since the oversea flight required in order to launch the missiles, mostly towards the Thames estuary, was short, often of less than half an hour's duration. Consequently, the bombs were comparatively well-aimed, as disclosed on the first of two maps prepared by the British Intelligence Service mentioned earlier. Accurate navigation of the aeroplane to the launching point for the pre-set missile was obviously a vital ingredient of success.

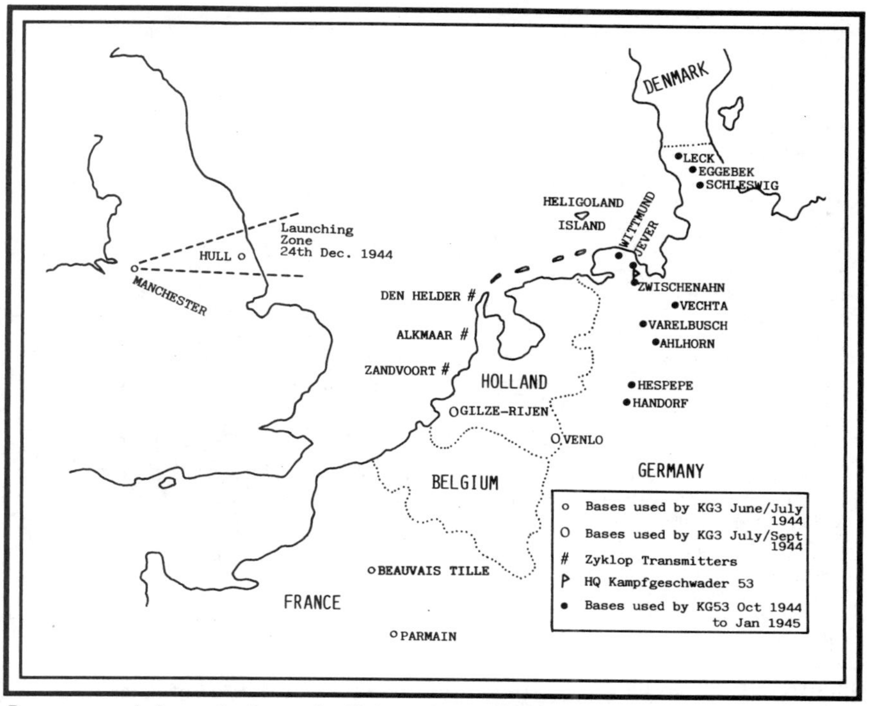

Bases used by air-launch V-1 units, 1944/45

The good visibility of the summer nights gave way to the less favourable weather of autumn and, finally, to the difficulties of winter night operations. As the weather deteriorated, and as the launching zone was pushed further northwards up the East Anglian coast in order to outflank the defences, the accuracy of the aiming decreased and this came as a welcome relief to the British authorities.

In order to keep the air-launch operations going in the darkness and the worsening weather, the Luftwaffe introduced more sophisticated navigational aids for KG53. These had to be proof against 80 Wing of the RAF, who specialised in interference with enemy radio navigational signals. Three powerful radio transmitters, called Zyklops, were set up on the Dutch coast, one each at Den Helder, Zandvoort and Alkmaar. By flying over one of these previously designated, the Heinkels were able to pick up a signal to start the digital run-down on the Zahlwerk. The aircraft then followed a directional beam emitted by this Zyklop transmitter on its flight over the North Sea to the launching zone. During the Manchester raid, all the KG53 Heinkels taking part flew to Den Helder before setting course for the launching zone.

A V-1 missile being hand-pumped up to height on the hydraulic jack of its transporter trolley to engage the suspension lug of the bomb to the catches on the aircraft. Two of the buffer arms against which the fuselage of the bomb rested can be seen in this photograph

In an endeavour to increase the accuracy of launching, a special pathfinder unit, I/Kampfgeschwader 66, equipped with Ju88 and Ju188 aircraft, co-operated with KG53. The pathfinder Junkers used the most up-to-date navigational aids. Each of its three Staffeln had a different type of navigational equipment. Two of these arrived at Handorf airfield on 18th October. They were equipped with Truhe 2, and with Bomber Benito. The third remained at Dedelsdorf awaiting fitting out with the more advanced Egon Luftkurier equipment. After preliminary trials, I/KG66 was ordered to state its requirement for "Schwan" radio navigational marker buoys on 6th November. On 11th, 12th and 23rd Nov-

Some air-launched V-1s had a blue/grey mottle sprayed on to them at the airfield where they were loaded on to the carrier aircraft, and this is very evident here

ember, German anti-aircraft units on the Dutch coast were warned of impending coastal crossings by the Ju88's and 188's along routes frequented by the Heinkels of KG53. The experiments were not, however, pushed to any fruitful conclusion and by December KG66 was involved in minelaying operations in the Scheldt estuary and in other operations connected with the Ardennes offensive.

Aircraft of KG3 or KG53 crossed their own territory at about 300 metres (1,000 feet) of height and dropped down to 100 metres (300 feet) for the oversea flight in an attempt to avoid detection by British coastal radar and to frustrate patrolling nightfighters. For similar reasons, the launching zone was kept some 40 or 50 miles out from the English coast. On approaching the launching zone, a fact indicated by the figure 1000 appearing in the Zahlwerk window, a rapid climb was made to 500 or 600 metres (1,700 or 2,000 feet), during which the speed of the aeroplane fell to about 110 mph. This was built up again during the level aiming run to at least the minimum launching speed for the V-1, about 150 mph. During the run, the course of the aircraft was also adjusted to match approximately that intended for the missile, whose own compass would attend to any final adjustment.

When the missile was launched, it fell about 100 metres (300 feet), before gathering sufficient speed for level flight. During this launching period leading up to the departure of the bomb, about seven minutes in total, the Heinkel was at its most vulnerable to nightfighter attack. Hans Hoehler, then flying with 7 Staffel, described it as a feeling of sitting in a lighted omnibus. He underwent the dreadful experience of suffering a nightfighter attack immediately after the launching of his aircraft's bomb at Manchester. The observer of the machine was killed and Hans Hoehler wounded in the neck. The remaining crew managed, nevertheless, to fly the damaged machine home to Leck and to crash-land there, concussing the already wounded Hoehler.

Nor were nightfighters the only hazard at this juncture; there is at least one known case of a V-1 stubbornly refusing to part from its parent aircraft and, with motor running, plunging both machines into the blackness of the North Sea.

The conduct of air-launch operations was carried out in an atmosphere of the utmost secrecy. The Geschwader code letters 5K for KG3 and A1 for KG53, never appeared on the usual fuselage side position on the aircraft. This secrecy extended to crew training, each member being gravely warned not to discuss any aspect of it outside the training area on threat of the direst penalties. To enhance the security, the aircrew lived away from the airfield. Typically, the crews of 8/KG3 based at Ahlhorn were billeted in the village of Grossenkneten, some 5 miles distant. The Staffelkapitan had his office in the village and practically all aircrew activities and recreations took place away from the airfield. Dry dinghy drill, for example, was held in the yard of the village inn. The crews were only taken to the airfield in darkness. Briefing took place there, but was short and fragmented. Written instructions were handed to observers and wireless operators and a talk was given on weather conditions likely to be encountered. The pilot, bomb-aimer and air gunner of the crew, therefore, knew little more than it was necessary for the other two members of the crew to tell them in the course of the operation.

V-1 Works No.707737 on a trolley in front of the camouflaged armourers' hangar at a KG53 airfield. The wings are stowed alongside the body of the missile, to be fitted immediately after loading on to the carrier aircraft

The aircraft were maintained and prepared by the ground-crew and stood ready for the aircrew to board on arrival. The specialist work of preparing and loading the missiles was carried out by V-1 armourers from the erstwhile Flak Regiment 155(W), by now re-formed and redesignated the 5 Flak Division.

The procedure of 9 KG53 ground-crews on Eggebek airfield prior to an air-launch operation was probably typical. In the afternoon the missiles were brought by field railway from their dispersal area in a wood to the preparation hangar. There they were assembled, fuelled and the various settings and re-adjustments made to the robot controls. The prepared V-1's were then slung under the Heinkels which had been brought by groundcrew on to an apron outside the armourers' hangar. Only after the fuselage of the flying bomb had been attached to the aeroplane could the wings of the missile be finally bolted on. A tricycle-wheeled hydraulic trolley jack was used to push the V-1 under the wing root of the Heinkel and raise it into position. This could only be done by approaching from the leading edge and the trolley only just cleared the starboard undercarriage leg. Thus the wings of the V-1 had to be off during this loading operation. The aircraft were first taken by the groundcrew from their dispersal points to the refuelling bays in pairs, and then towed to the V-1 loading apron. When so armed, they were taken to a position close to the runway and parked to await the arrival of their aircrew.

V-1 prepared for attachment to a camouflaged Heinkel. The missile wings were fitted last

Immediately after briefing, the crew boarded the waiting aircraft. The settings received by the observer were entered into the Zahlwerk boxes while engines were started and warmed up. On given signals at minute intervals, the aircraft took off. Runway lighting, including a static light at the end of the runway, was turned on for the takeoff. The overland course to the coast usually had a number of legs, each turning point being marked by a two-letter morse signal from previously laid-out, low-powered visual beacons. In a typical flight out of Ahlhorn on 7th October, the course lay north of Arnhem and out over the then Zuider See.

Over the sea the pilot used his FuGe 101 radio altimeter occasionally to check his barometric altimeter, but excessive use of any radio transmission equipment was severely discouraged.

Once set in motion, the digital counter in the Zahlwerk ran progressively down to zero. When it had dropped to 1000 the observer, who was also the navigator, instructed the pilot to climb to launch height. When the number 25 appeared on the Zahlwerk dial the bomb-aimer, checked by the observer, started up the pulse jet of the missile. Simultaneously, the robot's control system switched itself on. Finally, on the counter dropping to zero, the bomb-aimer released the missile and its departure was apparent to the crew from the cessation of its vibration and noise. There was also a noticeable lift of the aeroplane as it was relieved of the weight. Only ten seconds elapsed between the start up of the pulse jet and release of the bomb. The V-1 was then either seen to rush forward on its flight path as it picked up speed or, if the launch failed, it disappeared from view to splash down into the sea.

Apart from a glimpse as they boarded the aircraft, practically nothing could be seen of the missile by the crew during the flight to the launching zone. The relief felt on separation from such an onerous burden must, however, have been profound.

Three Heinkels of 3 Staffel 3/KG53 with flying bombs loaded ready for flight to a launching zone some 40 miles off the east coast of England. The low-level flight and pitch darkness lessened the risk of nightfighter interception but proved extremely hazardous

Chapter 5

STRETCHED DEFENCES

At the start of the flying bomb campaign, General Sir Frederick Pile, General Officer Commanding Anti-Aircraft Command, sought the approval of Air Marshal Roderic Hill, commanding the overall Air Defence of Great Britain, for a very large withdrawal of guns from the less exposed parts of the country. He proposed to withdraw almost all the regular anti-aircraft units from west of a line eventually drawn from the Humber to the Solent. This move enabled him to strengthen the defences of London and the South East by 204 heavy guns. A similar move of barrage balloon units was also ordered. Such anti-aircraft defences as remained in the North West were largely manned by Home Guard Anti-Aircraft units with a small cadre of regular gunners.

The risk to the districts denuded of defences was considered acceptable in view of the limited range of the V-1 and the spent state of the conventional German bomber force in the west. Further guns were found for the South East by selectively thinning out other defences. The end of June 1944 therefore saw a very large part of the nation's home defence anti-aircraft forces concentrated to defend the approaches to London from the Continental flying bomb launching sites. Unfortunately the shedding of the anti-aircraft defences of the North West region did not end there.

Typical of the Home Guard Anti-Aircraft units which remained available for duty was 103 Anti-Aircraft Battery, located in Alexandra Park in Manchester, now the YMCA sports ground off Princess Parkway. This was equipped with Z Battery rocket launchers, 64 of them, each capable of firing two rockets at once. Each launcher had a crew of two men who loaded, set and aimed the rockets on orders from central control. The unit was formed on 26th July 1942 and was actually setting up its equipment on 10th August, at the time when the last anti-aircraft fire was heard in the Manchester area. Intensive training was undertaken, including live firing practices at Freshfields, near Southport. The Home Guard personnel manned the battery at night by rotation, a small regular cadre looking after things during the day whilst the Home Guardsmen were at their civilian employment.

The siting in Alexandra Park was suited to the original intention of driving off piloted bombers from the city area. To counter a robot plane attack from the east, suitable open moorland sites were available within a few miles of the city. The possibility of V-1 launchings at targets in the North of England was obvious to Air Marshal Hill and General Pile by October. The large number of industrial areas of military importance in the region, within range of air-launched flying bombs from off the East Coast, created a potentially attractive target for the enemy. The gradual extension northwards of launchings up the coast of East Anglia pointed the way. By the time of an ADGB conference held on 21st September to consolidate a comprehensive plan for the defence, there was good evidence that the flying bombs aimed at London had crossed the coast even further north than Harwich. However, it was not considered justifiable to protect the whole East Anglian coast right round to the Wash. An order to deploy guns north of Harwich was, however, issued on the early morning of 22nd September.

There was an even more significant straw in the wind. At 4.13am on 19th September, a single air-launched flying bomb came inland over Skegness. Its engine cut out over mid-Lincolnshire, and it exploded at Tan Vats, near Metheringham, only 8 miles south-east of Lincoln. It blew a crater 6 feet deep and 25 across in a potato field. The missile had flown for 30 miles overland on a westerly course.

Despite the threat of V-1 attack, all the hard work put into 103 Home Guard AA Battery came to naught, for it was "stood down" on the last day of October 1944; indeed, the final stand down parade took place as late as 3rd December. Before the end of the month, T C Baskerville, in civilian life a scientific research officer, and a keen member of 103 Battery, wrote in his diary for 24th December:

"The attack of V weapons on the north has caused rather a stir among the northerners, who in my opinion, had thought themselves free from any further enemy action. It was to me very galling not to be manning our Home Guard Anti-Aircraft guns. Not an AA gun was going off. It was as at the beginning of the war when the enemy had the sky to himself. During two and a half years on the guns we saw no action and now we are stood down the enemy planes come back."

Barrage balloon personnel gather wreckage of V-1s brought down over southern England in the summer of 1944
(Imperial War Museum. Copyright reserved)

General Pile's concentration of anti-aircraft gun and barrage balloon defences in the South East was reorganised in July in the light of experience of countering the V-1 so far gained. The barrage balloons were gathered into a zone close to the southern outskirts of London, whilst all the guns were moved into a "gun belt" on the coast. This arrangement proved so successful as to suggest itself for extension to other parts of the coast which later needed protection from the incursion of air-launched flying bombs.

The barrage balloon screen was of peculiar advantage as a long stop in the protection of a highly concentrated, built-up area such as London, within a narrow band of approach. Its value disappeared and it was, in effect, discarded after the end of August. The gun belt idea concentrated on the coast, however, remained highly advantageous. The guns were by now able to score such spectacular results that a high proportion of incoming flying bombs could be shot down harmlessly into the sea.

The early air-launched V-1 operation led to the extension of the original Hastings to Ramsgate gun belt into a "gun box", placed to allow for continuous engagement of flying bombs as they flew up the Thames estuary. This box was bounded by an imaginary line joining Rochester, Whitstable, Clacton and Chelmsford. Standing fighter patrols were flown over the mouth of the river forward of the gun box. After the Germans regrouped and recommenced launchings towards dawn on 16th September, they attempted to outflank the gun defences further and the missiles came inland more to the north and east, over the Essex and Suffolk coasts. When this became obvious, Hill and Pile made the decision to extend the gun defences further northwards by adding a "gun strip" along the coast from the left flank of the "box", where it joined the coast at Clacton, progressively up to Great Yarmouth. The necessary guns for this were obtained by thinning out the "strip" made largely redundant by the Allied capture of the Pas De Calais launching sites.

The protective shield of these narrow coastal cordons of anti-aircraft batteries, only 5,000 yards deep, then extended more or less the whole way round the coast from Hastings to Yarmouth. By the middle of October, 498 heavy and 609 light anti-aircraft guns were deployed in the box and strip alone.

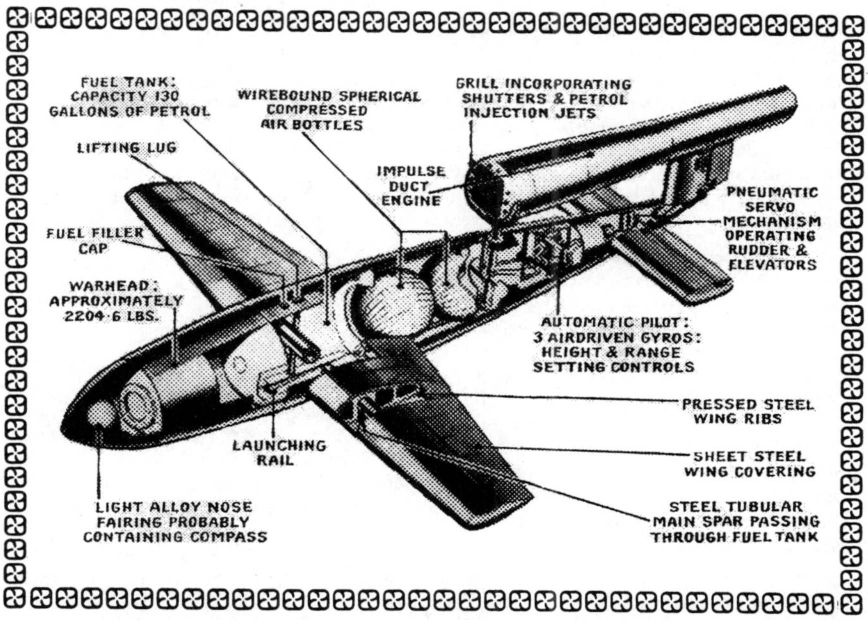

The British authorities were able to prepare this sectional drawing of the V-1 from wreckage found in June 1944

Those missiles air-launched to fly in over the flat lands of Essex and Suffolk came in very low and were thus subject to a minimum of exposure time to the individual guns which were brought to bear on them. At its extreme, this situation is illustrated by the experience of a Royal Observer Corps man, whose post was at Aldeburgh on the Suffolk coast and close to one of the AA gunsites. At 8.24 on the evening of 5th November, he reported to his headquarters at Colchester that a flying bomb had passed within ten feet of his post with its motor still running. It had torn through some of the tents used by the gunners, setting them alight. As ammunition was stacked in the area and the warhead of the V-1 did not appear to have exploded, he, very coolly in the circumstances, sought permission to evacuate his post. It was subsequently discovered that the warhead had indeed failed to explode and had been split into fragments when the flying bomb actually hit one of the guns. The fire caused was due to spillage from the ruptured fuel tank of the bomb ignited by the residual flame of its engine.

ADGB intelligence notes available in October put the range of the air-launch Heinkel bombers at 400 miles, to which was added the range of the V-1 itself, 150 miles. It was agreed that the time had come for preparation of a contingency plan for deploying anti-aircraft guns to counter possible incursions of flying bombs north of the Wash. On 2nd November, No.5 AA Group, then in the South East, was ordered to carry out a detailed reconnaissance for deployment into a zone from Skegness to Whitby. While the reconnaissance was being carried out by officers down to Battery Commander level, the Group Commander, General W W Green, ordered the guns and their platforms to be disposed so that the intended sites could be manned rapidly in the section between Hornsea and Mablethorpe. This was considered to be the enemy's most likely entry point in the event of attack upon the northern industrial areas.

The planning for the move included anticipatory arrangements for the billeting of troops and even involved the prior setting-up of some gun platforms. This greatly facilitated the movement into the new zone when it finally came about. Officers and NCO's selected for the move were all either experienced in, or were given specialist training in, the technicalities of shooting down flying bombs.

On 19th November, General Pile's plan for the new extension was complete. The stretch of coast between Skegness and Whitby was to be designated the "Gun Fringe" and the plan was submitted to the General Staff for approval, with a request for authority to commence the redeployment at discretion. It provided for the use of 59 batteries, including seven already in the Humber area defences. Bomber Command, who had a great concentration of airfields in use, both in Lincolnshire and Yorkshire, wanted to limit the Fringe as much as possible to avoid what they saw as a hazard to their own aircraft. The General Staff, therefore, delayed its decision and despite repeated requests from Anti-Aircraft Command, permission to make the move into the Fringe was not forthcoming.

At 4.40 on the morning of 18th December an air-launched V-1 fell in a field at Glebe Farm, Skeffington, ten miles east of Leicester, making a crater 10 feet deep. It caused damage to outbuildings of the farm and to a cottage. One person was slightly injured. Another missile fell in Rutland, about a mile east of Stocken Hall, then used as a depot for phosgene poisonous gas bombs.

In the early hours of Sunday 24th December, the feared attack on the North finally came. Approximately 50 Heinkels launched flying bombs from off the coast between Skegness and Bridlington; 31 of them came inland.

General Pile was given immediate authority to move 60 heavy anti-aircraft guns into the Fringe, followed two days later by four light anti-aircraft sections. This was a more powerful force than mere numbers would suggest. Enormous strides were being made in the science of anti-aircraft gunnery at that time and some of the new equipment which had been introduced accompanied the guns into the Fringe. Anti-aircraft radar equipment, the SCR 584, together with No.10 BTL predictors, greatly improved the performance of the guns. They plotted the flight of the bomb, automatically pointing the barrel of the gun towards the point where the flying bomb would be at the moment when the shell fired from the gun rose to the bomb's flight-path. The SCR 584 radar sets gave better readings in coastal regions away from any interference which ground contours tended to cause. No balloon barrage was ever contemplated for any of the coastal gun localities, as balloon cables, too, tended to interfere with the anti-aircraft radar.

Just as important as the radar and the predictors was yet another invention, brought into use in significant numbers only just before the opening of the flying bomb campaign. This was a new type of shell fuse invented in Britain, but developed in the USA. Nick-named "Bonzo", it was a variable time or proximity fuse, able to explode its shell in the optimum position if this passed near to the target. After firing from the gun, the proximity fuse began to transmit signals from a small radio. As it neared the target these signals were echoed and a receiver in the fuse thereby sensed the presence of the target and exploded the shell. These developments together went a long way towards the ultimate technical defeat of the flying bomb. By 12th August, General Pile was already able to claim that the average number of anti-aircraft shells fired per V-1 destroyed had fallen to under 100, and that the best batteries were shooting down a V-1 for every 40 rounds fired. As experience increased, this performance was much improved upon. A conservative estimate of the effect of the new equipment put a factor of 5 on the improved performance. Its full potential in the battle against the flying bomb was finally realised in the critical defence of Antwerp, concerning which Field Marshal Montgomery wrote on 12th April 1945: "The gunners have, during the last three months, raised the percentage of V-1 kills from 65% to 97%".

One good reason for deploying the guns along the coast in a thin band pointing seawards was the possible failure of the self-destruct devices fitted to the shells. Those fired out to sea did not even need self-destruct devices in most cases.

The V-1, whose course was straight and level, was in theory an ideal target for the

The confident tone of this newspaper report heralded the introduction of a British and American secret weapon to combat the flying bombs. The proximity fuse for anti-aircraft shells was brought into service just in time to counter the V-1.

BOMB SECRET OUT?

"You know, Smithers, I sometimes wonder if this flypaper idea of the General's isn't the answer to the whole thing."

THE Germans' new pilotless aeroplane — the robot bomb—beat us by a few days.

For over a year some of our best scientists have been working in a key Government aircraft department to find the answer to this "secret weapon"

They had all but got the answer when the bombs began to arrive. Now work is going on at even greater pressure, though the indications are that the Germans are unable to launch these bombs on a really big scale.

Attacks went on at intervals all yesterday afternoon, following a night of raids. But the number of planes arriving was much fewer than on Friday and they had a hot reception.

Next stage of our defence will involve powerful radio transmitters mounted on lorries or railway wagons.

Experiments to beat the new device have been going on in a remote part of the countryside, and yesterday General Sir Frederick Pile, Commander-in-Chief of the A.A. Command, was on a gun-site watching our new methods.

As Captain Harold Balfour, Under - Secretary for Air, said yesterday: "We shall soon get the better of this weapon.

HOW THE BOMB WORKS—PAGE 3.

AA gunner. But being small, fast and of a strong steel construction, it could survive many near misses. Further, it took the gunners some time to master the complexities of the new equipment. In the early stages of its use, unfamiliarity was the cause of many complaints which only patient tuition overcame.

Barrage balloons, whose original purpose had been to keep enemy piloted aircraft high and thus reduce the effectiveness of bombing, were not a deterrent to flying bombs. They were not deployed to protect the northern towns against this mode of attack, partly because of the hazard to our own aircraft.

The move into the Fringe took place over the Christmas period in severe winter weather. On 25th and 26th December, progress was held up by a thick fog which was followed by a hard frost so that it was difficult to dig in the additional portable gun platforms known as pile mattresses. There were heavy snowfalls on the Yorkshire coast which interfered with road movements and two sites were actually isolated for three days. Despite all the difficulties, the thorough preliminary planning paid off well and the initial move was more or less complete by 29th December.

The country south of the Humber, being low-lying and marshy with few and poor roads, posed the greatest difficulty. A substantial number of bridges needed strengthening before it was safe to use them for heavy loads, particularly the portable gun platforms. The Fringe was gradually strengthened until, by the second week of March, 152 heavy and 16 light guns were in it and ready for action. To man these required the personnel of 6 male and 13 mixed batteries. The majority of this force was found by redeploying weapons available from No.5 AA Group in the South East. Very few guns were withdrawn from the batteries on the East Anglian coast. There, throughout the air-launched V-1 attacks on London of December and January, the strength of the defences remained fairly constant.

Out of the original plan for the Fringe, submitted in November, only three of the five sectors designated were actually occupied. These were the central ones based on Sector Headquarters at Bridlington, Humber and Louth. Those intended to be based on Scarborough to the north and Alford to the south remained unoccupied.

Because Bomber Command opposed the extension of the Fringe into the Bridlington sector, this did not become operational before mid-February. It placed guns as far northward as Flamborough Head. The personnel to man it were obtained by redeploying two batteries from the Humber and bringing in six units from the south.

Dangerous as it appeared, the blow struck by KG53 at the North, with Manchester as its aiming point, did not employ the full available strength of the Geschwader. On 20th December, a date for which figures are available, I Gruppe had 40 Heinkels on strength (25 serviceable), II Gruppe had 36 (29 serviceable) and III Gruppe 39 (24 serviceable). There was always a number of machines unserviceable for various reasons. Since II Gruppe was still out of operation as a result of the series of accidents in mid-December previously mentioned, the burden of this attack fell on I and III Gruppen and effectively called out their whole serviceable strength.

The prospect of a sustained attack was more theoretical than real, however. There were factors acting to limit severely the scale of operations. By far the most important of these was the acute and increasing shortage of aviation grade fuel. During December this became so serious that there was never more fuel available than was required for the next operation. No stocks were held on any regular basis at airfields. KG53 was a victim of a deliberate Allied policy of defence through attack on the German oil industry. In May, Allied bombers started to concentrate their overwhelming power on to German oil industry targets. So successful were they that production of aviation spirit declined from 175,000 T in April to 156,000 T in May, 53,000 T in June, 29,000 T in July, 12,000 T in August and, according to the German Armaments Minister, whose figures these are, production was unlikely to exceed 10,000-15,000 T in September. In the event, no aviation fuel at all was produced between 10th and 16th September, and only 7,000 T for the whole of the month. The offensive then slackened, allowing production to rise to 18,000 T in October and 39,000 T in November. It then fell rapidly, ceasing altogether in February. The high stocks, which exceeded 500,000 T in June 1944, were consumed steadily and reached near exhaustion early in 1945. Thus was the whole Luftwaffe gradually reduced to impotence.

A barrage balloon in Alexandra Park, Manchester. These were all moved south when the V-1 campaign started in June 1944. Their cables brought down many flying bombs

The maker's photograph of a De Havilland Mosquito Mk1 from which the various nightfighter versions used to combat the air-launched V-1 attacks were developed

The inadequacy of aircrew training, amongst the reasons for which was fuel shortage, was also making itself felt. The personnel of KG53, both air and ground crew, were brought together around widely dispersed airfields in conditions of the utmost secrecy, with little opportunity to build up any real sense of identity. In particular, there was a lack of interchange of vital experience. All this was to result in a high accident rate and poor co-ordination. The temporary cessation of operations by II Gruppe as a result of accidents has already been noted. Sometimes accident occurrences became known to the British, who made frequent reference in their intelligence reports about this time to the low morale of aircrew in KG3 and later KG53. One such discovery happened on the afternoon of 6th October, when the pilot, bomb-aimer and gunner of a Heinkel were picked up from the sea, fifty miles off the Norfolk coast. On the previous night their aircraft, while engaged in an air-launch operation, had lost oil pressure in its starboard engine. The pilot had jettisoned the V-1 but height could not be maintained and he was forced to "ditch" in the sea. The wireless operator died in the crash and the observer drifted off in his one-man dinghy and was never seen again. German air/sea rescue services had ceased to function by this time.

Allied pressure limited air-launch activities from the very beginning. In the first half of June, III KG3 completed its air-launch training at Peenemunde and moved to two airfields in northern France; Beauvais-Tille and Parmain l'Isle Adam to the north-east of Paris. On 16th June, before operations had properly commenced, American bombers, by chance, dropped 54 tons of bombs at Beauvais-Tille and destroyed a number of Heinkels. From mid-September regular photo-reconnaissance was flown over what were considered to be the principal airfields then used by the air-launch units in north-west Germany: Varelbusch, Zwischenahn, Ahlhorn and (Munster) Handorf. On 23rd September, 23 aircraft of Bomber Command attacked Handorf. Five attacks were made on these airfields during the first week of October by RAF and USAAF machines and 217 tonnes of bombs were dropped. Air Marshal Sir Arthur Harris, the Chief of Bomber Command, did not, however, consider that airfields were suitable targets for his heavy bombers and showed less willingness to co-operate than Air Marshal Hill would have wished. A major bombing operation against the airfields by US Liberators, planned for 12th October, was cancelled on account of poor weather conditions. By the middle of October, the Chiefs of Staff thought the British home defences sufficient to contain the air-launched V-1. No further strategic bombing effort was, therefore, diverted to these airfields.

An armourer setting the autopilot of V-1 Works No.707737 for its predetermined course and height. Once set, it was essential that the carrier aircraft navigated accurately to the zone from which the autopilot setting had been zeroed

Air Marshal Hill was anxious that his night fighters harass and shoot down as many of the air launch Heinkels as possible in order further to demoralise aircrew "known to be already finding launching operations no little strain". Night intruder missions continually harassed KG53 airfields but few successes of a material nature were achieved. One experienced intruder crew had flown twenty missions against the airfields by November without any successes. On 5th January one Heinkel was destroyed on the ground and another damaged in one rare intruder success. The two main bases chosen for the anti-air-launch nightfighter operations were Coltishall, eight miles north east of Norwich, and Church Fenton fourteen miles east of Leeds.

Interception of the carrier

Heinkels by radar involved three stages. The long-range Chain Home Stations gave early warning to Ground Control Interception radar, whose range was about 50 miles. The latter "vectored" individual night-fighters, who then picked up the Heinkels on AI (Airborne Interception) radar at about 1.5 miles range. The system was both helped and hindered by circumstances. Sometimes Allied radio interception picked up and decoded German messages indicating continental coastal crossings by the Heinkels or their pathfinders. Conversely, the bombs were often launched outside or only briefly within GCI range. At best, the GCI stations would only hold the raider for about seven minutes, during which time it rose from 300 to 1,700 feet and carried out its launch.

The first two KG53 launch aircraft shot down fell on 25th September; one to 409 and one to 25 Squadron. Thereafter Mosquitos of 25, 68, 125 and 456 Squadrons scored further victories as shown in the table. These proved extremely difficult to gain, even hazardous for the fighter. The AI air-to-air radar did not work well at low altitude. Visual sighting was also exceedingly difficult. The dark green and black matt upper surfaces of the bombers were daubed with patches of matt blue-grey and this proved a most effective camouflage against a darkened seascape. Flame dampers shrouded the engine exhausts. A defensive tactic used by the Heinkels when seen was to take advantage of the low flying speed of which they were capable. The pilot turned into the path of the nightfighter and sought escape by melting into the background low over the sea, away from both radar and visual detection. Owing to the high stalling speed of the Mosquito fighter, there was a tendency to overshoot the target. If too low a flying speed was reached in an effort to keep the gunsights on the bomber, the Mosquito crew risked stalling into the sea without hope of recovery from so low an altitude. Near to the stalling speed, even the recoil of firing its guns could cause the fighter to fall out of the sky. There was little hope of successful engagement, therefore, except during the brief period of the launching run. Illustrative of the problems faced by the Mosquito crews was the account given to C F Rawnsley by one crew which had just landed after a frustrated interception. *"We were close behind him for*

Interceptions of Heinkel III carrier aircraft by RAF nightfighters

September 1944	
Night of 24/25	409 Sqdn claimed a HeIII destroyed over North Sea. 25 Sqdn claimed two HeIII's damaged. Another four were contacted on this night, but were lost.
Night of 28/29	25 Sqdn claimed two HeIII's destroyed over North Sea.
October 1944	
Night of 5/6	25 Sqdn claimed a HeIII destroyed over North Sea. 307 Sqdn contacted two HeIII's, but contact lost.
Night of 6/7	25 Sqdn claimed a HeIII destroyed and two contacted, but finally lost.
Night of 7/8	25 Sqdn aircraft fired at HeIII but result unknown. Mosquito failed to return. HeIII contacted, but lost.
Night of 9/10	307 Sqdn contacted a HeIII, but had to break off contact when guns failed to fire
Night of 12/13	25 Sqdn contacted a HeIII and obtained visual, but contact was eventually lost. 307 Sqdn aircraft reported contact with HeIII, but Mosquito failed to return.
Night of 16/17	68 Sqdn made contact with a HeIII, but finally lost it.
Night of 25/26	125 Sqdn claimed one HeIII destroyed (2/KG53).
Night of 30/31	125 Sqdn claimed one HeIII destroyed (4/KG53) and one HeIII probably destroyed.
November 1944	
Night of 4/5	FTDS claimed one HeIII destroyed (4/KG53) and one contacted, but lost.
Night of 5/6	68 Sqdn claimed one HeIII destroyed (5 or 6/KG53) and one HeIII contacted, but finally lost
Night of 9/10	25 Sqdn claimed one HeIII destroyed (3/KG53)
Night of 10/11	25 Sqdn claimed one HeIII destroyed (4/KG53) 68 Sqdn claimed one HeIII destroyed (4/KG53) and one HeIII as probably destroyed. 125 Sqdn claimed one HeIII as probably destroyed
Night of 14/15	68 Sqdn made four contacts on Heinkel carrier aircraft, but without result.
Night of 19/20	456 Sqdn claimed one HeIII destroyed (5/KG53)
Night of 22/23	456 Sqdn contacted a HeIII, but finally lost it.
Night of 23/24	456 Sqdn contacted a HeIII, but without any success.
Night of 24/25	456 Sqdn claimed one HeIII destroyed (1/KG53)
December 1944	
Night of 13/14	68 Sqdn made contact with a HeIII, but eventually lost it.
Night of 17/18	68 Sqdn contacted a HeIII, but without any success. 125 Sqdn claimed one HeIII as probably destroyed
Night of 22/23	125 Sqdn claimed one HeIII as destroyed (7/KG53)
Night of 23/24	68 Sqdn claimed one HeIII as destroyed (7/KG53)
January 1945	
Night of 4/5	68 Sqdn made two separate contacts on HeIII's, but lost both.
Night of 6/7	68 Sqdn aircraft contacted one HeIII, but nothing further known. Mosquito failed to return
Night of 12/13	68 Sqdn made two contacts with HeIII's, but both were lost.

Note: Up to the middle of October 1944, all the air-launchings were carried out by the III/KG3. Casualty returns are not available for this unit.

Confirmed destroyed	16
Claimed destroyed	4
Claimed damaged	2
Contacted, but eventually Lost	29
	51

(Compiled by Frank Leyland)

miles," the navigator said, "but he was so close to the water we couldn't get at him without swimming. I had to watch the altimeter as well as the radar, while the skipper tried to spot him."

"Didn't he pull up to launch the bomb?" asked Rawnsley.

"I'll say he did," the pilot broke in, "and as I spotted him, I could see the thing hanging under the wing, but before I could fire someone lit the blue touchpaper. There was a dirty great flash as the bomb went whooshing off on its way and for the next five minutes I could see 'Sweet Fanny'. I had to pull up in case I hit the deck and that was that."

(See "Night Fighter" by C F Rawnsley & R Wright. Collins)

A manufacturer's photograph of the Bristol Beaufighter from which various nightfighter versions were developed. Their low speed was an advantage in combating the KG53 Heinkels

Because the Mosquitos were proving too fast to cope with the tactics used by the German air-launch unit, it was suggested that outmoded Beaufighters of slower speed be brought back into operation. The Experimental Fighter Interception Development Squadron had several available Beaufighters already equipped with radar. These had a much lower stalling speed and it was decided to send Squadron Leader Jeremy Howard-Williams with one of them to Coltishall for an operational trial. On 4th November, the Hopton Ground Control Interception coastal radar station guided the Beaufighter crew to a point at which they were able to pick up a target Heinkel on their own radar. Squadron Leader Howard-Williams described it as looking like a sinister black crow. He shot it down in flames, bomb and all, leaving a brief patch of burning fuel on the surface of the sea. Shortly afterwards he saw a flying bomb being launched by another Heinkel two miles away. Cautious of overshooting, he approached it slowly. It was clearly illuminated by the pulse jet flame of its missile and was seen to bank away to port after parting with the V-1. Although contact was obtained on the AI radar, this was lost again as the radar beam reflections from the bomber began to merge with those from the surface of the sea, which it again hugged for the return flight. (See "Night Intruder" by J Howard-Williams. David and Charles 1976.) The Beaufighter had proved a success, but it was evidently not considered practicable to re-adopt it on a larger scale for these operations.

The indicator and receiver units of the AI Mark VIII radar in the two-seated cockpit of a Mosquito nightfighter. On the shrouded screen the navigator was able to follow the "blips" which enabled him to guide the pilot until he could see the faint outline of an air-launching Heinkel

Later a naval frigate, HMS Caicos, and also a Wellington bomber fitted with air-to-surface radar, were tried in an attempt to guide the Mosquitos to the low-flying and elusive Heinkels, but by that time the air-launch campaign was almost over. In all, some 24 of the air-launch aircraft were attacked by nightfighters on operational patrols and, of these, 16 were destroyed, one of them by Mosquito TA389 of 68 Squadron during the Manchester raid. Another ten fell to naval guns. During the whole air-launch V-1 operation a total of 77 machines of KG3 or KG53 were lost from all causes.

On 28th December Adolf Hitler wrote: *"We are causing continual disturbances to the English industrial regions through these flying bombs"* - a reference to a situation which, however exaggerated, was almost over.

After the Manchester raid, missiles were launched by KG53 on only four more nights and all were aimed at London. The Oldham incident was the last really major V-1 catastrophe and in three more incidents only were the casualty lists to show more than ten people killed. On 14th January the very last air-launched V-1 to fall in Britain hit Hornsey on the northern outskirts of London. British air intelligence reported in February that "some general limitation, probably shortage of aviation fuel" had "brought about a continued relative inactivity" in KG53 operations.

As a result of the Manchester raid, Central Fighter Establishment of the RAF at Wittering in Cambridgeshire moved 68 Squadron, equipped with Mosquito marks XVII and XIX, to Wittering on 2nd February 1945. The belated move was an attempt to cover the direct approach to the industrial Midlands.

In reality, the air-launch campaign was over. A brief renewal of ramp launchings from Holland in March, using longer-range bombs against London, was a last, almost ineffectual gesture.

In the Gun Fringe, General Pile's gunners waited in vain for V-1 targets. Some of them saw a brief moment of action against a renewal of Luftwaffe piloted aircraft raids in March. These were directed mainly at RAF Bomber Command airfields along the length of the East Coast. The gunners were able then to show off their potential value and there is, perhaps, a touch of irony in these gunners being so well placed to protect the localities of airfields whose commanders had raised such strong objections to their deployment into the Gun Fringe. General Pile's struggle to stretch his defences as far north as Flamborough Head was not entirely wasted.

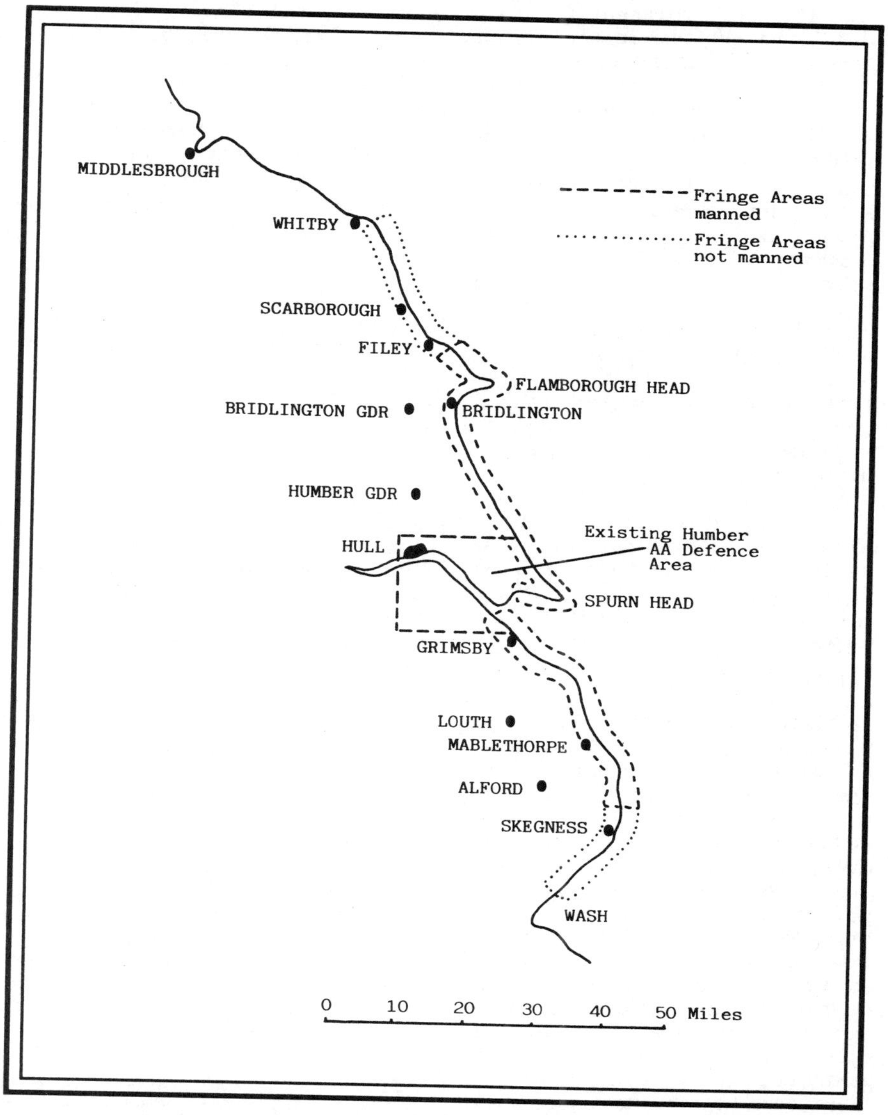

Gun defence development following the raid of 24th December 1944

Chapter 6

'DOODLEBUGS - ALL UNDER THE BED'

Amongst the several RAF coastal radar stations to receive warning of enemy air activity, gleaned from intercepted wireless messages, in the early hours of 24th December was that at Benacre Park, Wrentham, Lowestoft. On duty there as an NCO Radar Operator was a Manchester man, F J Stott of Fallowfield.

His shift had begun at 11.00pm and throughout the night he scanned the sky over the North Sea with an 11° lobe from the 240 feet high radar mast. The equipment was part of the protective Chain Home System, with a range approximating to 100 miles, according to weather conditions. Orders were given by the WAAF Supervisor to "inch" the search into a narrow band to the north and soon the "blips" of raiding KG53 Heinkels appeared on the screen in a position some 40 or 50 miles off the coast of Lincolnshire. Operator Stott watched as the aircraft climbed to height and saw the "blip" jump and divide into two as the flying bomb separated from the carrier and sped towards the coast. Some six or seven aircraft appeared within range and in each case the bomber dropped back towards the sea after despatching its load. Time and location were reported back to Group Headquarters at Stanmore. The operator was well aware that he was observing the launching of flying bombs at the North of England, perhaps at his home town of Manchester.

Launchings were witnessed by several trawler crews and later reported. Skipper Gorringe of the steam trawler Gurth, fishing 7 miles south-east of the Dowsing Spar, saw one successful and four or five failed launchings. Skipper Miller of the Scouter, fishing 14 miles north-east of 62B buoy, saw up to 18 launchings, mainly to the north-west of his position. He saw at least one bomb go into the sea.

The launching took place over a period of about an hour, roughly between 5.00am and 6.00am. None of the V-1s took more than half an hour to complete its independent journey.

As the bombs rattled inland, seven were seen and engaged by the Humber gun defences, but these had no success in shooting any of them down and no overland attempt was made at fighter interception. Two did, however, fall in the Humber area shortly after crossing the coast. One of these came down at Willerby on the western outskirts of Hull, the other in the estuary itself.

Only one bomb fell within the Manchester Special Civil Defence Area. This approached over Stockport. It crossed Didsbury Road and the River Mersey at East Didsbury. Just beyond Cheadle it commenced its dive and the engine cut out. Taking a turn to the right, it again crossed the Mersey and at 5.30 exploded in a field of Brussels sprouts fifty yards from the north bank of the river. This point is within the river bend, two hundred yards south of what was then Underbank Farm at the bottom of Millgate Lane, Parrs Wood. To the south of the river at this point lies Gatley.

Flying bombs often behaved in an erratic manner once the spoilers at the tailplane had put them into their final dive. In this case the bomb dived steeply to the right. Often they would continue on course in either a steep or a shallow dive; sometimes the bomb would even spiral down. Densely populated Wythenshawe lay directly in the flight path of this missile and was probably saved from a disaster by the fortuitous turn. Fortunately it exploded in one of the most open areas in the whole of the central part of Manchester and there were no serious casualties. A crater 9 feet deep with a diameter of 18 feet was blown in the soft loamy soil of the river valley. Because of the softness of the ground, and consequently deeper penetration of the bomb, blast damage in the vicinity was less severe than it might otherwise have been, though windows were broken over a wide area including Fog Lane, Burnage; School Lane, Didsbury, and Altrincham Road, Northenden. Many shop windows were shattered in the old Sharston shopping area.

Mr R Heywood saw the bomb plainly from Warsop Avenue, Sharston, and particularly noticed the amount of light emitted by its pulse jet. In Altrincham Road, Mr F Robinson heard it as "a dull, pulsating roar" that "seemed to shake the bed" as he "realised that it was a doodlebug", and in the brief silence that followed the cut-out of its motor "waited for the bang".

In Branksome Road, Heaton Mersey, 16-year-old Joseph Critchley wakened to the combined noises of the air raid siren on the roof of Moorhouses Mill, Cheadle Heath, and the approaching V-1. He leapt out of bed in the darkness at the very moment that the missile seemed to be overhead. With alarm, he heard the engine cut. In his attempt at a quick exit from his unlit bedroom, he ran into the half-open door and "saw stars" - an indirect casualty of the bomb. Whilst still recovering from the accident he heard the explosion of a second missile, the one which fell slightly later at Adswood, Stockport.

On the sounding of the "All Clear", a general discussion of the raid was joined amongst the neighbours and he dared to suggest that it was flying bombs that had caused the explosions. His listeners were unimpressed - "Don't talk silly, lad, doodlebugs can't get this far" - was a typical rebuke reflecting the general disbelief.

Clifford Blackburn was a cowman for Joe Woodbridge, who then farmed Underbank. Clifford took rapid shelter on hearing the missile and when he reported for work slightly late, Joe and his sons were already at the crater. Some damage had been done to the roof and windows of the farm and a lean-to greenhouse was practically destroyed. Close to the crater ran the old road to Gatley Ford, along which a column of the invading Jacobite army had passed in 1745, at the last land invasion of England.

The site has been much changed since 1944. Filling and levelling have converted it into playing fields and Underbank Farm has been pulled down to make way for Underbank House. An electricity pylon now stands very close to where the crater once was. The M63 motorway passes within a few hundred yards on the opposite side of the river.

One of the missiles passed almost over the city centre, causing alarm amongst the early traders at the old Smithfield vegetable market.

A nurse working at Booth Hall Hospital at the time noticed the speed with which those of the little London evacuee patients who were awake, recognised the sound of the doodlebug. One, a nine-year-old boy, sat bolt upright in bed, listening, and then shouted, "Doodlebugs! All under the bed!" She was amazed to see many of them quickly take cover. He then shouted, "Duck under the bed, Nurse!"

Even war work was interrupted. 23-year-old Jessie Green was at work on the 10.00pm-6.00am shift at Connolly's Cable Works at Delaunays Road, Blackley, when the Red Alert sounded just before 5.30.

Instructions were that in this

V-1 damage at Worsley

event all machinery was to be stopped and the workers were to evacuate the building and proceed to the air raid shelter. Jessie was asked to step outside a set of double doors to collect some milk which they intended to take with them to the shelter. As she pushed open the door to get at the milk, she heard the noise of a flying bomb in a direction over her left shoulder, and looking south towards Crumpsall Hospital and Cheetham Hill, she saw what she described as a "moving object with a flame pulsing from it, low in the sky".

She knew, from having seen the cinema newsreels, that it was a flying bomb with its engine running, and quickly ducked back into the workshop where the noisy machinery had just stopped, and from which people were making their way towards the shelter.

Only she had seen the missile and she found it difficult to convince her friends in the shelter that she really had seen a V-1 over Manchester.

None of them had heard it explode at Worsley, nor did they hear the explosion of the Radcliffe bomb 20 minutes later.

The "Red Alert" system had been devised to avoid lost production time in factories engaged in important war work. Special warning was given when there was immediate danger from enemy air activity so that production was able to continue during normal air raid alerts. Thus the effect of air raiding was minimised in this regard, at some additional risk to war workers.

Sixteen-year-old Alex King had spent the previous afternoon at the Tatler News Theatre in Manchester and, like Wilfred Swann, had seen, amongst the films, "Story of the V-1". Wakened from his bed at Emerson Street, Weaste, by the siren on the Weaste bus depot, he was dressed and at his front gate talking to neighbours when the same missile roared over Buile Hill Park to the north of him, travelling west at great speed. Despite the glow and noise of the pulse jet, he too had great difficulty in making the bystanders believe that they had witnessed the passage of a buzzbomb. Nobody seems to have thought of seeking shelter, though several distant explosions were heard.

The bomb impacted at 5.30, some 108 feet to the north-west and in the rear of a detached house, No.18 Woodstock Drive at Worsley. This house was demolished, together with its next door neighbour, No.16. Two other houses in Woodstock Drive were also severely damaged, and lesser damage was done over a wide area. Woodstock Drive, only one side of which was then built up, is on a gradient and the bomb exploded at the higher end. This somewhat lessened the blast effect at the more populated lower end near to Worsley Green. The air raid warning preceded the bomb by only five minutes.

The first Civil Defence rescue party, consisting of seven workers, arrived at 6.00am to find four people trapped in the damaged houses, which were of brick construction with slate roofs.

Mr Statham of No.18 was rescued, with head injuries, as was Mrs Arnsbury of No.20; Mr Arnsbury suffered a fractured leg and shock. A tragic fatality of the incident was the Arnsbury's 6-year-old son, who was sleeping downstairs because he was recovering from measles.

Shock and cuts were suffered by Miss Cowley of No.24, and she was taken, along with the other seriously injured, to Hope Hospital at Salford. Constance Cardwell of No.26 escaped with lesser injuries, but was given first aid.

At 8.32 it was reported, from the incident site to County Civil Defence Control at Preston, that a cylindrical object 10 feet long and a metal sphere with a tube attached had been found in the crater. Control warned that the sphere "should be handled with care as it may contain compressed air". The cylinder was probably the remains of the Argus jet duct and the sphere, which was later reported to be "crushed and therefore safe", was one of the compressed air bottles of the robot.

One and a half miles to the south-east, 16½-year-old Roy Davies was on night shift, working as a train recorder in the Patricroft signal box. He and the signalman were aware that there was an air raid alert in force. They heard a "put-put, like a motorcycle engine" and went to the end windows of the box facing towards Manchester. The engine sound "misfired" and then there was silence. Both signalman and train recorder made a dive for the "steel cabinet" - a Morrison shelter kept in the signal box for such emergencies. Surrounded by glass, they realised their danger. Cramped together inside this refuge, they saw the "flaring of light" emitted by the explosion and a moment later heard the bang. It seemed to them that this may have been enemy action directed at the Royal Ordnance Factory, only a few hundred yards away from the signal box in the direction of the explosion.

Almost due east of Woodstock Drive, close to where the Worsley Road crosses the East Lancashire Road, young Robert Cliff was wakened by the air raid siren and had some difficulty in convincing his parents, brother and sister that they should get up and go into their shelter. This was in the cellar, but approached through a lean-to greenhouse at the rear of the building. Before reaching it they heard the unusual engine noise which grew rapidly louder and then cut out when it seemed right overhead. Father and son saw the light of the pulse jet and then the robot itself, which they took for a conventional aircraft. As it fell it "disappeared behind some houses", followed by a flash and a bang.

This bomb appears to have entered into a steep dive after the spoilers activated, but to have maintained its directional course. On its flight it passed over Newton Heath, Collyhurst, Queens Road, Cheetham, Higher Broughton and Irlams o'th' Height. In order to clear the Pennines, the robot's altitude must have been set at not less than 2,000 feet above sea level. The high volume of noise and the amount of light emitted gave the impression to observers of a lower height. The relative altitude would, of course, be reduced by the height at which the observer was himself standing above sea level.

The implications of the V-1 incursion into the North were not lost on the better informed. Alex King's uncle was an ARP Warden in Weaste. When Alex mentioned to him that he was convinced that it was a flying bomb they had witnessed, his uncle, looking serious and pensive, said, "Yes, I'm afraid it was." They both wondered when, rather than if, a further attack would come.

Chapter 7

NOT THE PENDLETON FAULT

In Chapel Street, Tottington, on the site of property destroyed by a V-1, is a memorial garden given by Mr and Mrs S D Whitehead to the people of Tottington "in memory of those killed by a Flying Bomb on 24th December 1944" and a plaque in the garden records the names of the dead:

Nicholas Conway (50), a dye-works operative
Mary Conway (49), his wife
Elizabeth Draper (56), a shop assistant
James Dyson of Balwell, Nottinghamshire
Teresa Dyson, his wife
Ann Greenhalgh (75)
Bertha Greenhalgh

Mr and Mrs Dyson had arrived in the village only a few hours earlier to spend Christmas with Nurse M Rooney, the sister of Mrs Dyson. The nurse was on night duty when the bomb fell and returned to the remains of her home to help in the rescue work and in the location of the bodies of her sister and brother-in-law.

Mr George Ormerod, who owned and lived in the damaged property, was away on holiday in Lyme Regis for the Christmas period, a fact to which he may have owed his life.

The injured who were taken to Bury General Hospital included Miss Mary Conway, Mr and Mrs Dewhurst Greenhalgh, Mrs Midgeley, Miss Ethel Riley, Mrs Hodges, Mr Herbert Young, Miss N Hemes and Miss Ellen Barnes. Fortunately, none was very seriously injured.

Mr R Robinson, a porter in the hospital, started his duty at 6.00am on 24th December. He left his home in Fir Street, Bury, at 5.35 and, walking up the right hand side of Walmersley Road just before 5.50, saw a V-1 pass over from east to west.

Having served in the army in the London area during the summer, he was familiar with the V-1 and quickly took cover, face down in a small wood by the side of the road, and shortly heard the bang of its explosion.

As soon as he arrived at the hospital, the Night Sister on duty instructed him to waken the hospital doctors in expectation of the arrival of casualties. When these did arrive in various vehicles, he helped move them to the treatment rooms and wards. Dr Josephs took the precaution of telephoning Bury Town Mortuary to arrange for transfer of the dead from the small hospital mortuary which had only four spaces.

Back at the incident, a nearby school was used as a rest centre for the homeless, who were all later found accommodation by relatives or friends.

Two evacuee families from London lived in one of the destroyed houses; Mrs Barry with her three children and Mrs Anderson with her two. By good fortune they had all left the previous afternoon to return to London for the Christmas holiday.

It was 5.50 when the V-1 hit the row of terraced cottages in Chapel Street. Numbers 21 and 23 were totally destroyed and severe damage was done to the two adjacent properties. The frontage of these four houses, including a narrow passage between them, was some 55 feet. The houses were constructed of regular coursed stone rubble with slate roofs. They were of two storeys with cellars.

A shop in the same row and close by was practically destroyed and Chapel Street was blocked with debris. St Ann's Church, on the opposite side but standing back from the road, suffered from both blast damage and from flying debris. All the windows were blown in except for the one behind the altar, which, ironically, had only recently had its bomb protection cover removed because of the absence of air raiders for so long. The churchyard trees were littered with articles of clothing, bedding and furnishings blown there by the blast. The vicar, the Rev H W Rogers, held the Christmas services in the church despite the damage. He and the local constable, PC Lovatt, were two of the first rescuers on the scene after the V-1 exploded.

27 houses were seriously damaged, eight of them being rendered uninhabitable, and 350 suffered some degree of blast damage.

Within a few minutes, Rescue and Ambulance Services were on the scene and an incident post was set up at Tottington Police Station, from which the rescue operations were directed. Such was the size of the task before the rescuers, that nearly ten hours of digging and rubble removal were necessary before the body of the last of the victims was recovered. At 2.55 County Control was requested to relieve the rescue party then on duty and to send a mobile canteen. At 5.42 the Incident Control Officer was finally able to report that the rescue operations were completed, and that in addition to the six dead, a final total of fourteen people were injured. Fifty-three people had been made homeless. By the end of the day these had been found accommodation and the rest centre which had been set up was closed.

Clearing the debris of wrecked cottages in Chapel Street, Tottington

The remaining bombs which penetrated into Lancashire were fortunately less serious in their consequences.

At 5.54 a missile passing to the north of Oldham crossed the centre of Radcliffe, impacting half a mile beyond. It fell on waste land some 30 feet to the south-east of a works used by Prestolee Activities Ltd, and close to the bottom of School Street. A crater was blown 20 feet across and 4 feet deep. Severe damage was done to the works building, a brick structure 140 feet long by 40 feet wide with an asbestos sheet roof.

Prestolee Activities, a confusingly named company since it operated in Radcliffe rather than in nearby Prestolee, were makers and repairers of wooden lifeboats. Some 27 of these were damaged or destroyed in the explosion. The firm employed 12 people and is remembered for its profuse use of brass nails.

The works probably grew up on this site because of its proximity to the Bury branch of the Manchester, Bolton and Bury Canal, the nearby section of which remained open in 1944, though its adjacent section to the south-west had been closed and drained on the orders of the Ministry of Transport. This ran along the side of the Irwell Valley and otherwise would have constituted a far-reaching flood risk should an enemy bomb have breached its banking.

Happily, there were no casualties, but in addition to the wrecked works, a large number of wooden hutments and greenhouses of the numerous allotments in the vicinity suffered. The "Dig for Victory" campaign must have taken a minor setback.

The blast, wide-ranging as usual, affected other property over a radius of about a mile. Hundreds of windows in works, shops and houses in Radcliffe were blown out. The local YMCA and the club houses of two bowling clubs were damaged, and the nearby school lost most of its windows.

Left on her own whilst her husband was in the army, newly-married Mrs Payne lived in Buckley Street, less than a quarter of a mile from the explosion. Fortunately for her she had attended a wedding in Manchester on the Saturday afternoon and stayed over, to return to Radcliffe on the Sunday morning.

On the train she heard about the Radcliffe bomb and returned home to find her windows broken and front door so distorted by the blast that it had to be forced open.

The crater was not far from the base of a redundant, but strongly built, brick industrial chimney, and this must have been severely shaken by the concussion. Nevertheless, it remained intact. The blast effect to the westward was largely dissipated in the open ground between Radcliffe and Little Lever, over a mile away.

The site of the crater is today covered by a scrap metal business, one building of which follows the approximate line of the old Prestolee building, but is of smaller area.

At the very top of Plantation Road, Broadhead, Turton, is Red Earth Farm on the border of rough, high moorland. At 5.44 a missile fell on the moor some 200 yards south-east of the farm and on higher ground. A crater 13 feet wide and 6 feet deep was made. Very recently this rough ground has been turned into marginal grazing land, but the crater was until recently still to be seen. There were no casualties and damage was limited to the glass of the farm and some other buildings within the usual mile radius. It was not until 9.35 in the evening that any report of the Turton bomb reached County Civil Defence Control. This report was recorded in the message book there as "believed to be a V-1". It was noted that "fragments and paper cartridges containing pamphlets had also been found".

As recently as 1983 rusty fragments of this missile were still buried in and around the crater. The young son of the occupant of Red Earth Farm overheard the writer's enquiry about the crater and was able to identify it quite accurately, having played up on the moor, even though he had no previous knowledge of what it was.

At Brindle, in the rural district of Chorley, two cottages were wrecked, two farms seriously damaged and minor damage was done to 112 houses, 2 mills, 2 public houses, a chapel and a railway signal box. This flying bomb, the first of the raid to reach its destination, exploded at 5.28. It fell on farmland about 30 yards in the rear of Stonefield Cottages, a pair of semi-detatched dwellings at Milnthorpe, between Gregson Lane village and the railway level crossing near Hewn Gate Farm.

In one of the cottages Mr Joseph Bolton, a farmworker, was in the course of lighting the fire in his kitchen grate when the explosion brought down much of the house around him and trapped his wife and daughter in their beds under debris. Mr Bolton was trapped by one leg and foot, but fortunately managed to extricate himself, less one of his boots, and gallantly went to free his family. Luckily, all suffered only minor injuries. Trixie, their cat, was so scared that she fled and only came home again the following day.

In the next door cottage, Mrs A Hodgson lived alone. She too was covered in debris in her bedroom, but managed to free herself and was relieved to find her stairs still standing so that she was able to make her way to ground level, clad in her nightdress and cardigan. In the darkness, each step

Debris in St Ann's Churchyard, Tottington

down the damaged staircase was a hazard.

Mr G A Birtill, examining the crater next day for his local newspaper, found leaflets released by the bomb and drew the attention of the policemen to his unusual find. He was promptly relieved of them and, despite his pleadings to Superintendent Joe Cook at Chorley Police Station, was told that it was more than the Superintendent's job was worth to hand them back. Mr Birtill reported his find to the Press Association, but the story was held up for "security reasons".

The missile made a spiralling turn during its descent and came down on a hen coop containing 30 pullets. Only traces of straw, wood ,and the feathers of these birds were ever found.

Some confusion existed in the minds of the Civil Defence Personnel over the incident. The County Controller received report No.31 from Sub-Control F at 11.45am, purporting to cover two incidents:

"Chorley RDC reported two V-1 bombs
(1) - 164470 map ref. (the old pre-1947 map references were used) -houses and mill damaged by blast.
(2) - 068474 map ref. -2 houses demolished, namely Stonefield Cottages, Milnthorpe - no casualties. Police have matter in hand.
Chorley RDC standing by."

Both sets of damage were in fact caused by the same V-1, as was later apparent.

A girl staying with friends at Longridge, Preston, heard a "crash" and felt a slight shake in the night caused by the Brindle incident. When she mentioned it to her friends at breakfast, they told her that it was an earth tremor caused by the Pendleton Fault, a local geological feature which sometimes caused slight earthquakes.

The site is about two miles east of the present M6/M61 motorway intersection. Where Stonefield Cottages stood there is now an agricultural workshop built of corrugated sheeting. A stream runs through the uneven meadow in which the crater was made. There are now no signs left to indicate its exact position and some tipping in the intervening years may have changed the ground contours.

V-BOMB ATTACK

DAMAGE IN RURAL AREAS

IN a BOMB attack in the north of England, cottages twenty yards away from where a bomb fell, in a rural area, were wrecked but the occupants escaped unhurt.

Though the district is an isolated one, many windows were smashed, including those of a village church. Four hundred windows were shattered in one building, a mill.

The bomb scored almost a direct hit on a hen cabin, and one of the occupants of the nearby cottages, Mr. Joseph Bolton, a farm worker, was lighting the fire and preparing to go to work when "the whole place came down."

Chorley Guardian, 29/12/44

Lower Westall Farm stands to the left of the Haslingden to Blackburn road, three miles short of Blackburn and within the boundary of Oswaldtwistle. As the road takes a turn to the right, the track to the farm goes off up an incline to the left. To the right of the track and short of the farm is the sloping meadow in which a V-1 exploded at 6.00am. This was the last of the Lancashire flying bombs to fall. Local people had plenty of warning on this occasion, as the alert was sounded at 5.28. The crater blown was five feet deep and twelve across. Roof, door and window damage was done at Lower Westall and at nine other farms and cottages. A small pottery works also suffered damage, but there were fortunately no casualties resulting from this incident.

One local farmer, on the opposite side of the main road, remembers the sound of the bomb "as though it were running out of power". In the darkness and mist he was under the impression that the point of impact was much closer to him than, in fact, it was. Daylight revealed it to be nearly three-quarters of a mile away.

His barn doors and all his windows facing in the direction of the explosion were blown out and a number of fragments came down on and about the farm. He saw the larger remains of the missile being removed from the crater area by the authorities as he came back from delivering his milk.

Including Manchester within its boundary, the then administrative county of Lancashire had received eight of the bombs and by far the heaviest casualty list of any county involved in the raid.

V-1 crater at Red Earth Farm, Turton, photographed in 1983

Chapter 8

THE SECRET OF MACCLESFIELD FOREST

The spread of flying bombs actually crossing the Pennines was almost equally divided between those to the north and those to the south of Manchester. Cheshire received six and one wandered south into Shropshire.

Westwood Farm, Matley, Hyde, lay to the left of the A57 Hyde to Mottram road, 100 yards up a track opposite the New Inn. Today, the M67 motorway cuts across at the back of the house which has replaced the farm. To the left of the track leading to the farmyard a meadow occupied the space between the main road and a line of trees fronting the farmhouse.

Farmer Edwin Foulkes had brought his mother-in-law, Mrs Greenwood, over from Failsworth on the Friday afternoon. She was to spend Christmas with the family; Mr and Mrs Foulkes, their sixteen-year-old son and thirteen-year-old daughter, Betty.

It was Edwin Foulkes's habit to rise at 6.30 to do the morning milking. He was already awake at 6.25 when he briefly heard the unrecognised sound of the pulse jet followed, after a pause, by a tremendous shattering bang and the disintegration of everything around him. The bomb detonated in one of the trees opposite the front of the farmhouse. The whole floor of the bedroom collapsed to ground level.

Mother-in-law and son, both sleeping in rooms at the back of the house, were killed by the blast, but strangely, Mr and Mrs Foulkes, nearer to the explosion and exposed to its full force, escaped with serious injuries, as did their daughter. Mrs Foulkes suffered permanent eye damage and Edwin Foulkes carried a scar on his face from the injuries sustained until his recent death.

Six cattle in the shippon were either killed or fatally injured. The remaining sixteen were rescued after the shippon had been shored up. A roan pony used for the milkround, and to which its owner was much attached, was also destroyed because of its injuries.

Edwin Foulkes was born in 1896 and served in the Royal Flying Corps in the First World War. He never recovered his will to farm, and later built the present house on the site. The death of his son, the destruction of his farm and last, but not least, the looting of his property whilst he was in hospital in Hyde, recovering from his injuries, were altogether too much to bear.

The roof was torn off the New Inn on the opposite side of the main road and two cottages in the rear of the farm were also badly damaged.

Michael Moores in Dukinfield heard the explosion in the night but was unaware of any air raid in progress. In the morning news of the "buzz-bombs" rapidly spread by word of mouth and on Boxing Day he joined "hundreds of others" in walking to inspect the wreckage of Westwood Farm.

The Borough of Hyde Civil Defence Committee met on 3rd January to consider a report of the incident, but this meeting was held in secret and the report itself has since disappeared. The committee did, however, adopt certain of its recommendations: "That the First Aid Post in Union Street be retained" and "That two satisfactory vehicles be provided for the Civil Defence Rescue Service".

Garners Lane, Adswood, lies one and a quarter miles south of Stockport centre, off Bramhall Lane. At 5.30 a V-1 coming in from the west along the line of Garners Lane detonated on a bank of earth some 40 yards from the roadway at the rear of house number 87. Both 87 and 89, a bungalow, were wrecked, the frontage involved being 35 feet. Two adjacent houses suffered severe damage, some 50 were seriously affected by the blast and 300 suffered lesser damage. A crater 15 feet wide by 3 feet 6 inches deep was made. Both 87 and 89 were of similar construction to the other properties in the lane, brick-built dwellings with tiled roofs.

There was a disturbing casualty list – two men and two women seriously injured; one man, four women and fourteen children injured to a lesser degree. The one probable fatal casualty was reported in the "Deaths" column of the Stockport Express of Thursday 28th December:

"Mr Joseph Briscoe – On 24th December, 1944, aged 75, Mr Joseph Briscoe of 71, Garners Lane, Adswood, Stockport – the interment took place at Stockport Borough Cemetery on Wednesday 27th inst. following a private Church Service in St George's Church.

Although Mr Briscoe was seriously ill his death occurred under tragic circumstances in the early hours of Christmas Eve, doubtless hastened by an unexpected incident. Until his retirement fourteen years ago he was head of the well-known firm of public works contractors which bore his name."

Newspapers were always considered by the authorities to be a possible source of information to the enemy and for this reason there was a strict embargo on the linking of any

Westwood Farm before it was destroyed

place names with the reporting of air raid incidents, hence the guarded reference to the cause of Mr Briscoe's death.

Mr and Mrs William Etchells both received severe head wounds when number 87 was blasted. Mr Etchells was found by rescuers part buried under a jumble of the bricks and floor boards of his home. Both he and his wife were taken to hospital. Great care had always to be exercised in the rescue of buried casualties in order to prevent further injury or the collapse of unstable parts of the surrounding building. The specialist training of the Civil Defence rescue parties was directed to this end.

Others seriously injured were James and Freda Bourne. Mr Bourne had a badly fractured arm and his wife facial injuries. The first warden to reach the Bourne's home heard a voice shouting, "We're in the back bedroom". He entered the house through the smashed front door and by the light of his torch saw that the bedroom floor was sagging badly. Mrs Bourne, who was covered in debris to her waist, said, "Get the baby out," and her small daughter, Jennifer, was handed to someone outside.

The demolished bungalow, No.89, was owned by Mr and Mrs Gregg, who were away from home, but a friend, Mrs Squires, who was staying at the premises "for quietness", as she told a local newspaper reporter, was sleeping under a window on the side facing the explosion and remarkably escaped injury despite the full force of the blast. Next door stood another bungalow, from which the occupier, Mrs Higson, also escaped without injury.

A local Civil Defence Post Warden, C E Richardson, was about 200 yards away round a bend in the road when the explosion occurred. He saw the flash, which temporarily impeded his vision, and since it was very dark, he did not know where the bomb had landed. As his duties required, he set about searching his sector. Meanwhile, his daughter, who was a Civil Defence messenger, started helping in the rescue work.

Some adjacent houses were rendered uninhabitable. A great number of broken windows included all those in the local golf club house. From the centre of the V-1 explosion, blast waves normally radiate outwards in concentric circles like the ripples made by a pebble thrown into a pond. These waves are alternately of high and low pressure but the effect is capricious. Some houses close to the explosion escaped damage altogether, while others a considerable distance away lost windows and roof tiles. A row of shops on Bramhall Lane lost windows and suffered frontal damage from the blast.

An incident post was set up in one seriously damaged oak-panelled sitting room. Evidence of Christmas decorations still adorning the damaged rooms added a pathetic touch to the scene.

The Chief Constable of Stockport was also the Civil Defence Controller. Talking to newspaper reporters, he expressed his satisfaction with the ready response of Civil Defence workers to the emergency, occurring as it did after their duties had been relaxed under the stand down arrangements of the autumn. While regional services had been brought in to help, it was the local services - Wardens, rescue people and the ladies of the WVS who, coming back into harness so quickly, satisfied him so very much.

Not all the citizens perhaps behaved quite so gallantly. It was reported that there was one fractured leg and one badly lacerated face in another part of the town, suffered as people were "going to the shelters".

The danger of a panic amongst nervous people anxious to get under cover in the public shelters was very real and had been highlighted in London on 3rd March 1943. While people were taking cover at the Bethnal Green Tube shelter, a sudden firing by a local anti-aircraft battery had started a rush during which a woman fell, blocking a passage at the bottom of some stairs. Pushing began and soon people were piled one on top of another. In all, 173 people were crushed or suffocated to death. As early as the First World War fourteen people had died of suffocation under rather similar circumstances, at Bishopsgate on 28th January 1918.

A local American army unit sent a party of 150 soldiers to assist with the emergency property repairs at Adswood. They were particularly helpful in cutting up and fixing sheets of roofing felt, which had been somehow found from stocks, despite the large quantities sent south during the flying bomb attacks of the summer. Much of this they used as a temporary cover for the broken windows.

The Mayor of Stockport visited the site and said that his main concern was to get the slates back on to the less seriously damaged property so that it might be kept dry. The weather helped, as over the Christmas period it was dry but very frosty. Stockport's sixth and last wartime Christmas was reported to have been the coldest for 72 years.

The inevitable curiosity of the sightseers upset one warden, who told a reporter that he had been engaged in the rescue operations from the beginning of the incident and that, even before it was properly light,

Westwood Farm after its destruction by the V-1 on the night of 23rd/24th December 1944

people who were strangers to the district had come "nosing round just to see the damage". He said that it "went on all day and that not one of those he accosted offered accommodation for the homeless, or even asked what they could do to help," adding that there were plenty of people who needed help.

One early riser, Mr Clifford Bates, was walking along the road close to the explosion and survived to give his account. He was flung against a wall by the blast, but otherwise was unhurt. When he recovered himself, he went to the scene of the explosion to find wardens already there. He praised their promptitude and "marvellous work".

J R Kitchen of Heaviley, about a mile to the north-east, was woken by what he thought was a "two-stroke motorbike noise" and for a few seconds his bedroom was faintly lit by a strange glare, the light from the pulse jet. The noise stopped and a loud explosion followed. Soon afterwards he was called out by his employers, the Stockport Corporation Electricity Department. He and two colleagues were sent to cut off the electricity supply to a number of the damaged properties.

He particularly remembers the work of the Salvation Army, who had a mobile canteen on the site at a very early stage when it was most urgently required. Copious supplies of tea were much needed by rescuers working in the dust of the pulverised brickwork.

The reason for the unusual line of approach of the bomb to Garners Lane was that it spiralled during its dive, impacting at a moment when its direction of flight was exactly opposite to its original course.

Almost opposite the Setter Dog Inn at Walker Barn on the A537 Macclesfield to Buxton road, a lane runs off to the hamlet of Five Ashes. 300 yards beyond the hamlet, within the area known as Macclesfield Forest, a "fly" impacted at 6.10 between Crooked Yard Farm and Clough House Farm, blowing a crater 27 feet across and 5 feet deep in a sloping hillside pasture.

A few minor injuries resulted, only two requiring treatment, and some damage was done to neighbouring farms. Mr W Slack at Five Ashes, a boy in 1944, has good cause to remember the explosion since his bedroom faced towards it at a distance of about 400 yards. The windows were blown in, showering his room with splintered glass. Some others were blown out, complete with frames. One metal window frame was sprung out and carried across the yard by the blast. It was later put back and reglazed. It gives sound service still, none the worse for its travels. The house is of solid stone construction, but was so badly shaken that only in recent years has it been made truly watertight again.

Master Slack and his brother, Eddie, had a secret to keep. They picked up some of the leaflets and magazines which were strewn around by the bomb. Hearing that the police were gathering all these in, they decided that they wanted to keep a few, particularly of the magazines which had coloured covers. Some they buried close to the roots of a tree in the corner of a nearby field. These were never dug up again.

Nearer to the bomb crater by 150 yards was Crooked Yard Farm, Macclesfield Forest. This had much of its roof blown off and was covered with tarpaulins for several months, even beyond the end of the war in Europe four or five months later.

In 1983 Mr Slack helped the writer recover the remains of the tail pipe of the pulse jet of this V-1, partly crushed by the explosion and suffering from the consequences of rusting, but still complete and retaining its tail fin connecting fork. This had lain near the crater for many years and had then been taken to a local farm as a curiosity. It is now restored for exhibition purposes and at the time of writing is in the Manchester Air and Space Museum.

The Macclesfield Forest bomb fell three miles to the east of Macclesfield town; another fell three miles to the west, at

The author with the tailpipe of the V-1 which fell at Macclesfield Forest

Henbury. This exploded, fortunately without causing any casualties, in a hollow of a field about 30 yards out from the edge of Bluebell Wood. The wood skirts a lane leading to the A537 at Henbury village. On the opposite side of the lane lies Henbury Hall, standing in its own grounds well back from the wood. Nevertheless, it was the massive old trees of Bluebell Wood which largely shielded Henbury Hall, about 400 yards away, from anything more than window damage.

The V-1 struck at 6.05, making a crater 10 feet across and 4 feet deep. There are no signs of this crater in the field now, but a few fragments of the bomb were recently found with a metal detector. The trees in the wood probably have some fragments still buried in them, imbedded there by the blast.

The edge of the wood facing the impact site still clearly shows where the great boughs were torn off on the cold, dark December morning over forty years ago.

Further down the A537, near to where it passes through Ollerton village and two miles to the south-east of Knutsford, lies Manor Farm on Seven Sisters Lane. Mr Harold Calwood, who now farms it, remembers hearing the approaching pulse jet from the Macclesfield direction and the terrific explosion as it landed at 6.05, making a shallow crater in the meadow adjoining the western side of the farm. The tail pipe was blown right across Seven Sisters Lane and, striking a tree, came down in a field a hundred yards from the crater. A farmworker who had quarters in one of the outbuildings was very slightly injured; fortunately he was the only casualty. The roofing and windows of the farm suffered some damage. The siren in Knutsford had sounded the alert at 5.29; the "All Clear" followed at 6.40.

Two days after the bomb had fallen, leaflets released at its final dive were discovered still in their wrapping in a hedge half a mile to the south-east of the crater.

The clay soil was blackened by the explosive gases for a radius of about 20 feet from the centre of the crater. People living nearby described the noise of the pulse jet as "sounding like the throbbing of a motorcycle". Doors and windows shook in Knutsford. Souvenir hunters amongst the large number of visitors to the farm on the following day found fragments of twisted steel, part of the remains of the missile. As recently as 1982, the writer found one of the robot's shattered gyroscopes still buried a few feet from the crater. The position of the crater can still be located as a shallow bowl in the meadow a few yards from a semi-circular hedge that then bounded a pit.

The most westerly V-1 to fall exploded at 6.22 in a field belonging to Lower Farm, Kelsall, only seven miles from Chester and the tidal water of the west coast. A large crater, 20 feet across and 10 feet deep, was made in a remote field which abuts the land of Street Farm. Both farms stand to the left of the A556 Northwich to Chester Road.

Mr Booth, still resident in Kelsall village, was at that time employed at Street Farm and had just arrived for work. He was in the shippon with the farmer's son, Robert Tomlinson, and the farm foreman. They heard a sound coming from the east over Kelsall Hill. Looking out of the shippon door, the foreman saw the V-1 and said it was a plane on fire. The sound stopped and Robert Tomlinson suddenly realised that it could be a flying bomb. He called to the others to get down, and just as they did so it went off. The straw and "dust" from the shippon floor thrown up by the concussion "seemed to come up to meet them". There were no casualties but some damage was done to the roofs of the Street Farm buildings and one roof truss was broken. Many slates were taken off local roofs and windows were shattered. Leaflets were scattered over the area of the field and many were later taken as souvenirs, particularly by locally based American soldiers.

Some of the larger remnants of the bomb were later dumped into a pond surrounded by a spinney at the edge of the field.

The writer, who was then 13 and living near Northwich, remembers being woken by his father just after the noise of the missile on its way to Kelsall had passed. There was a period of anxious waiting and listening to see what else would happen. Then came the "All Clear" siren. News that it was a flying bomb that had disturbed the peace, and that it had fallen at Kelsall, eventually filtered back. News of the other flying bombs of that night was limited to a later discussion amongst relatives, telling that there had been another bomb at Adswood and one in Manchester.

The Shropshire "fly" was one of those which glided for a considerable distance. Its motor, according to the police, cut out over Stafford and its axis of flight, west by south-west, indicates that it came from the southern part of the launching zone, probably north-east of Skegness.

It exploded at 5.40, making a shallow crater 19 feet 7 inches by 17 feet and only 18 inches deep in a field at Middle Farm, Chetwynd, about a mile out of Newport. A gap 105 feet long was torn in a hawthorn hedge some 1,600 yards south west of

The author with the gyroscope of the Ollerton V-1

Aquelate Hall, then used as an army depot.

The crater was a few hundred yards behind buildings now used as riding stables, to the left of the A518 Stafford to Newport road. Sixteen plate glass windows were broken in the town and a considerable number of smaller windows, mainly in domestic property, were also broken.

In daylight, the police examined the site and reported:

"Details of Fly -
(1) Lettering on Fragments -VOR DEM STARTEN FERNEN. 701 431. (Note: This should read: Vor dem Starten entfernen -withdraw before launching - and was stencilled on the tail fin; a part of the preparation instructions for the groundcrew.)
(2) Jet propulsion tube in the crater, east end.
(3) Balloon cutters fitted to wings.
(4) Copies of V-1 POW Post
(5) Copies of Signal Magazine.
(6) Direction of flight west by south-west.
(7) Four photographs were taken."

Mr R J Marsh, whose father farmed the land in 1944 and who himself remembers the incident, thinks that the shallowness of the crater was due to the ground having been frozen hard at the time. This would have influenced the depth of the crater, as would the shallowness of the glide from Stafford and the considerable variation in sensitivity of the mass-produced electrical fuses fitted.

Chapter 9

STRAY BIRDS

Traditionally the Pennines formed a physical and psychological barrier between the Lancastrian and his eastern neighbours. Huddersfield and Halifax, over the high and remote moors, seemed further away than the distance measured on the map showed them to be. Some of this remoteness remained until the coming of the motorways, and it seemed a protection against enemy activity from the east. The path of the buzzbombs lay over the high moors, which as a consequence received their quota of the missiles.

The Macclesfield Forest bomb was the westernmost to fall on the hill chain, but it was closely followed by two others, where the Pennines merge into the Peak District.

The spa town of Buxton in Derbyshire lies in a bowl in the hills and forms a major trans-Peak road junction. It had two lucky escapes from "flys" within five minutes of each other. The alert sounded at 5.26 and lasted until 6.36.

At 5.45 the first of the two buzzbombs expended itself high up on the moorland of Black Edge, four miles to the north of the town and half a mile from the A624 Buxton to Chapel-en-le-Frith road, above the village of Doveholes. The crater, originally 33 feet across and four feet six inches deep, is now a shallow pond but identifiable some 150 yards beyond the sharp contour of the Edge itself. The writer recovered the remains of one of the bomb's three gyroscopes from this crater in 1987.

The only casualty attributed to the missile was one sheep which it killed. Disturbance was caused in an Italian POW Camp about a mile away, and a few windows were broken in the camp and in six civilian houses.

The second missile passed over Buxton and exploded at 5.50, less than two miles to the west of the town. The trackbed of a dismantled mineral railway runs under the A54 Buxton to Congleton road and passes north to the mouth of a bricked up tunnel entrance. Parallel to this is the wooded slope of Burbage Edge, with the moor stretching behind it. The trackbed emerges again into sloping ground, and only 200 yards beyond and to the left of the tunnel exit lies the crater. It was originally 35 feet across and 5 feet deep, but it has been eroded with the passage of time. Again, a single sheep was reported killed, but there were no human casualties; nor was any damage done.

Over the Yorkshire border, two missiles fell in the Pennine area. The first came down at 6.00am in a field only 30 yards from the rear of Little Tootill Farm, Hubberton Green, two miles to the south-west of Sowerby Bridge. Serious damage was done to the old stone-built farmhouse, which has now been rebuilt as a pleasant dwelling house and renamed Little Tootill Cottage.

Tootill End Farm, some 300 yards to the north-east, also suffered superficial damage and fragments of the missile fell in the hamlet of Hubberton Green. The crater measured 26 feet across and was 4 feet deep. Leaflets, many still wrapped in their paper tube packages, were found in the area. Local legend has it that the lady occupant of Little Tootill Farm, whilst not listed amongst the casualties of the raid, was "never the same again" after her experience. She died, it is said, as an indirect result some time later. A second bomb was heard to pass over the district on its way into Lancashire. It was noted at the time that typical fragments of the V-1 were painted red on the inside surface (doubtless the primer coat) and that no cable cutters were found from the remains of the wings.

Beyond the Woodhead reservoirs, the A616 for Sheffield leaves the A628 at Flouch Inn, short of Penistone. From near this junction a footpath crosses the Little Don river at the end of Langsett Reservoir and leads up on to Midhope Moor. Some four miles along, the path passes through Cut Gate and on under the slopes of Margery Hill. At 5.45 a "fly" blew a wide crater 4 feet deep and 40 feet across in the rough moor to the right of the footpath and on the lower slope of the hill. The crater is now filled with water, but has recently been re-identified from fragments of the bomb found around it.

Since it is a five mile walk from the road, the crater was not approached until Christmas Day 1944. It was then described as showing marks of a white deposit and as having fumes still rising from it. Both phenomena are more likely to be attributable to local ground and perhaps weather conditions. So far from habitation, no damage was caused and even the sheep seem to have escaped fatalities, for none was reported.

The title of this book embraces those buzzbombs which crossed or fell upon the Pennines. The story, however, would be incomplete without a mention of the bombs which were launched by KG53 and crossed the coast that night but failed to reach the "backbone of England".

One was a very stray bird indeed. Either intentionally or more likely through a malfunction of its compass, it flew north-west and fell on the Tudhoe Colliery cricket field at Charles Road, Tudhoe, a mile north of Spennymoor in County Durham. The vicarage and 17 other houses were badly damaged; some 70 suffered minor damage, as did the church. One man and three women were seriously injured.

A small crater 6 feet across and 3 feet deep was officially reported, the time of impact being given as 6.05. The warning was sounded at 6.01 and the "All Clear" at 6.38.

Something of a mystery surrounds this incident. According to one account, the bomb came down at a shallow angle and struck frozen ground behind the vicarage, then skidded for some distance before reaching the cricket field. According to this account, petrol, which spilled from the ruptured fuel tank of the missile, caught fire. The remains of the bomb are said to have been removed to the main police station in Durham and to have remained there until at least 1946.

Further south, a trail of shortfalls marked the course of the missiles from the east coast to the eastern slopes of the Pennines. Any technical failure or serious loss of power resulted in a shortfall or a stray bomb. The longer the range at which the robot was set to fly, the greater the chance of its failure. In the case of the Manchester raid, the missiles were set to fly near to their maximum range.

At Dumbsteeples Farm, Grange Moor, Kirkburton, five miles short of Huddersfield, extensive superficial damage was done to 120 small houses at 5.35 when a crater 25 feet across and 4 feet deep was blasted out of a field. This was only 120 yards from the nearest of the houses. Fortunately there were no casualties.

Only 60 feet from the old course of the Great North Road at Rossington, 5 miles south of Doncaster, a ploughed field was cratered at 6.00am. This hole was 4 feet deep and 20 feet across, and was only 100 yards to the north east of Rossington Hall lodge gates. These were dislodged and telephone wires beside the main road were brought down. In 1944 most telephone wires were carried on poles beside public roads and the consequent temporary disruption to communication of damage at an arterial route could be extensive. Other superficial damage was done to a farmhouse, but no casualties resulted.

Military objectives, particularly airfields, anywhere within easy range of the east coast during the war were always liable to air attacks. The most unlikely of these ever to occur was probably that made when one of the doodlebugs landed near to the north-west boundary of Barmby Moor airfield at Pocklington, ten miles south-east of York. The bomb fell only 25 yards from a four-engined RAF Halifax bomber of 102 Squadron, parked at dispersal. A crater 12 feet across and only a foot deep was made in a grass field close to a gravel pit and just 50 yards from the nearest house. This suffered substantial damage and eight other houses superficial damage. A concrete builder's workshop close by was demolished. It is known that the Halifax bomber was severely damaged, but whether it was treated as a write-off or repaired subsequently is not clear.

The site is now within the grounds of Stobbs Wholesalers Warehouse, off Back Lane in Barmby Moor village, between the back of the manager's house and the old airfield perimeter. Six of the damaged houses were in Back Lane. One of the residents described how she and her husband heard the sound of the buzzbomb's approach and, recognising that it spelled danger, got down on the bedroom floor for cover before the explosion occurred. Because of the damage done to their home, they were evacuated to live with neighbours for three weeks whilst repairs were carried out.

Nine miles to the south-east, on South Cliffe Common, another V-1 fell at Bunnyhill Farm. It hit a ditch bank 1,000 yards south-east of the farm buildings, to which no damage was done other than the breaking of a few windows. Other buildings suffered minor damage, mainly in the hamlets of North and South Cliffe.

The most easterly V-bomb of all to fall on land during the raid came down in the outskirts of Hull itself, at Willerby. This made a crater 12 feet across and a shallow 18 inches in depth in a grass field only 150 yards from Springhead Water Pumping Station, which suffered some damage. The nearest house, 130 yards away from the crater, was amongst several to suffer superficial damage. Windows were broken over a wide area in the western part of Hull and one lady was slightly injured.

Brian Hansby witnessed the descent of this missile from outside his parents' house in Ridgeway Road. This was at one end of a block and separated from the next block by a passage 10 feet wide. Flying very low, with its engine still running and a long exhaust flame trailing, it narrowly missed the chimneys. It appeared to follow the course of the passage and impacted just beyond a track leading from Kingston Road to the pumping station. Fortunately, Brian's father threw them both to the ground before the bomb struck about six or seven hundred yards away. They went over to the crater and saw the two spherical air bottles and the pulse duct lying in the vicinity. The missile had passed low over the centre of Hull, which thus had a lucky escape.

A doodlebug flying up the Humber estuary was seen by a Civil Defence worker, Mr K May of Winterton. From Ferriby Sluice on the south bank, west of Barton-on-Humber, he watched as it started a spiralling descent and, with its engine still running, plunged into the water to the east of Reads

The shallow, water-filled crater made by a V-1 on Midhope Moor (Photo: Dr Alan Curry)

		Time a.m.
	CHESHIRE 6 BOMBS	
1	KELSALL	6.22
2	OLLERTON	6.05
3	HENBURY	5.45
4	MACCLESFIELD FOREST	6.10
5	STOCKPORT	5.30
6	HYDE	6.25
	SHROPSHIRE 1 BOMB	
7	**NEWPORT**	5.40
	LANCASHIRE 8 BOMBS	
8	OLDHAM	5.50
9	DIDSBURY MANCHESTER	5.30
10	WORSLEY	5.30
11	RADCLIFFE	5.56
12	TOTTINGTON	5.50
13	TURTON	5.44
14	OSWALDTWISTLE	6.00
15	**BRINDLE**	5.28
	DERBYSHIRE 3 BOMBS	
16	BUXTON	5.50
17	CHAPEL-EN-LE-FRITH	5.45
18	**BEIGHTON**	5.40
	NOTTINGHAMSHIRE 1 BOMB	
19	STURTON-LE-STEEPLE	5.45
	LINCOLNSHIRE 2 BOMBS	
20	REDBOURNE BRIGG	5.40
21	**EPWORTH**	5.48
	YORKSHIRE 7 BOMBS	
22	ROSSINGTON	6.00
23	GRANGE MOOR	5.35
24	MIDHOPE MOOR	5.45
25	SOWERBY BRIDGE	6.00
26	WILLERBY	5.45
27	SOUTH CLIFF	—
28	**BARMBY POCKLINGTON**	5.50
	COUNTY DURHAM 1 BOMB	
29	**SPENNYMORE**	6.05
	NORTHAMPTONSHIRE 1 BOMB	
30	WOODFORD	5.30
	IN HUMBER ESTUARY 1 BOMB	
31	READS ISLAND	—

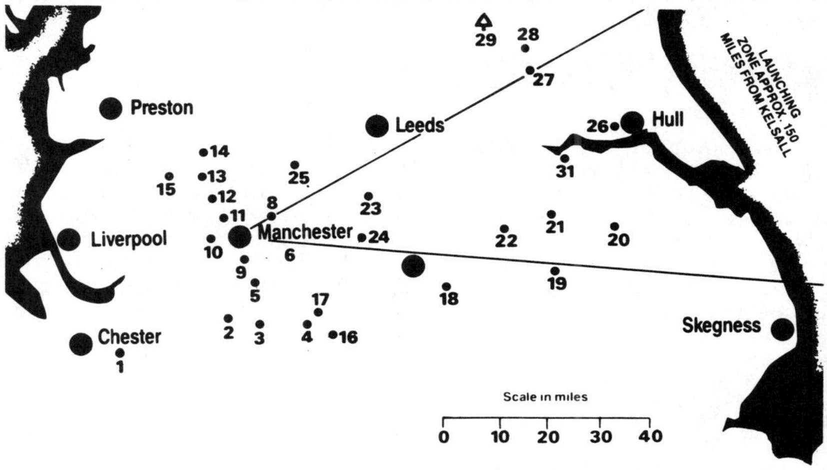

The air-launched flying bomb attack resulted in 42 people being killed, 51 seriously injured and 58 were less seriously injured

Island, apparently without exploding. It seems likely that either it had been unknowingly damaged earlier by guns of the Humber defences, or that it developed a defect which brought it down. It still lies somewhere in the river mud. The distance which it had flown since launching may have been insufficient for the airlog to have activated the electrical fuses which would, in that case, still be at a safe setting. A gradual spiral descent, particularly into water or soft mud, would be insufficient to detonate the mechanical inertia bolt fuse which the weapon had as a reserve.

South of the Humber in Lincolnshire, one shortfall detonated at 5.40 in a stubble field 350 yards east of Woofham Hill Farm, Redbourne, 1.8 miles away from the A15 south of Brigg. Some fabric damage was done to the farm and to the windows of its neighbour, Paradise Farm. There were luckily no casualties.

Another came down at 5.48 near to Epworth on the Isle of Axholme, north of Gainsborough. It fell on arable land, making a crater 12 feet across and 4 feet deep in Clarks' beetfield, 150 yards south east of a mill. Some damage was done to the mill and to the ceilings and windows of some 40 houses. A preliminary Civil Defence report indicated that the warhead of this bomb had failed to detonate, but the presence of the crater and the obvious blast effects gave the lie to this and it was later corrected.

Close by was a Royal Observer Corps post manned by two men, Mr G Revell and Mr E Lindley, and connected by telephone to their control centre at York. The observers correctly identified the V-1, Revell calling out over the telephone the code words "Diver, Diver," followed by an alarmed "Look out! The d..... engine has stopped and it's coming down. If we go off the air you'll know that we have had it." York heard the explosion over the telephone line, but luckily neither of the observers was injured, nor were there any other casualties.

By a strange irony, ten weeks afterwards, on 4th March, the only ROC observer to be killed on duty in the North died less than 20 miles away when his car was struck by a crashing Ju88 intruder between Welton and Hackthorn.

Only one missile fell in Nottinghamshire, at 5.45, 2 miles west of the River Trent, which forms the county boundary there with Lincolnshire. The crater, 4 feet deep and 18 feet across, was on ploughed land off Littleborough Road, Sturton-le-Steeple, in the Retford rural district. Fragments were scattered over a radius of 300 yards and part of a thorn hedge was torn up. Little other damage was done and there were no casualties.

At 5.40 a bomb impacted at Beighton, six miles to the south-east of Sheffield, the crater, 3 feet 6 inches deep and 14 feet across, being in grassland. Some damage was done to Field Farm just to the north of it, and windows were broken in some 150 buildings in Killamarsh and district, but no casualties were caused.

The remaining robot strayed into Northamptonshire and at 5.30 exploded in an open field at Woodford, a mile to the south-west of Thrapston. The crater, 5 feet deep and 20 across, was 450 yards to the north-west of Grange Farm, to which some superficial damage was done. Glass and tile damage occurred to other buildings in the neighbourhood but there were no casualties. A small brown paper parcel containing copies of the "Signal" magazine was recovered from near the crater.

There was a spread of 170 miles between the northernmost bomb at Spennymoor and the southernmost at Woodford, but excluding these two and the one at Newport, the spread was approximately 40 miles north to south and overland east to west about 100 miles.

Of the 31 missiles which crossed the coast, 19 fell within 30 miles of the centre of Manchester. 15, approximately half the total, fell within a 20-mile radius and six of these fell within 10 miles. Of the estimated total of 50 missiles launched, 38% had failed to cross the coast, but 12% had fallen within a radius of 10 miles of the target, 30% within 20 miles and 38% within 30 miles.

Chapter 10

INSIDIOUS PAPER

From late August 1944, some flying bombs arriving over Southern England had been equipped with a device enabling them to discharge a small payload of propaganda leaflets. This was activated by the airlog of the robot on the same signal as that triggering the dive spoilers.

The leaflets were packed in pasteboard cylinders or paper cartons. Many of these packages were found intact and typically were cylinders $2\frac{1}{2}$ inches in diameter and about 18 inches long. A small explosive charge similar to a spoiler detonator was fired. This both impelled the packages of leaflets from the bomb and ignited a touchpaper on each package. As the touchpapers burned away they freed the leaflets from the packages as these descended. The individual sheets were intended to drift down and spread themselves over the locality. The first of them were found on 29th August and came from missiles which fell at Sittingbourne, at Dartford and on Romney marshes.

The early V-1 leaflets were illustrated and of two types - an 11 x 7 inch size folded sheet headed "This is an experiment, let's try it - W Churchill" and the other, of much smaller size, about 4 inches by 2 inches, was headed, "Do you like this? You do? You may not in a few months' time". Both showed photographs of air raid victims in Germany. The first was intended to evoke sympathy, the second, which was sometimes included as a slip addition to the first, as a threat of further retaliatory action.

Two other leaflet types followed in November 1944; one folded as a small newspaper and headed, "The Other Side", aimed to put a German view of events to its readers. The other was headed, "A Splendid Decision", and claimed to be of the same title and on the same theme as a paper written by J M Speight, a former Principal Secretary to the British Air Ministry, in which he was said to acknowledge that Adolf Hitler only undertook the bombing of British civilian targets reluctantly after the RAF had commenced bombing German civilian targets.

Occasionally the release mechanism failed to operate completely and the leaflets stayed with the bomb to be blown to fragments. This happened when the bomb impacted prematurely, as at the Willerby incident, where only two leaflets were found in a readable condition. On other occasions, some already cited, the packages were thrown clear but failed to break open and were later picked up intact. Instructions for manually opening the packages were printed in English upon them, and involved tearing off a strip of paper.

The Ministry of Home Security ordered that all enemy leaflets be treated with the suspicion that they might be associated with bacteriological warfare and that they be gathered in by the police whenever possible. This was made much easier, of course, in cases where the packages had not broken open.

Samples were to be sent for laboratory examination. An extract from "Consolidated Instructions on the disposal of Unexploded Missiles", issued in January 1944, contained the following paragraph:-

"Propaganda Leaflets - Copies of enemy propaganda leaflets are sent by the police to a

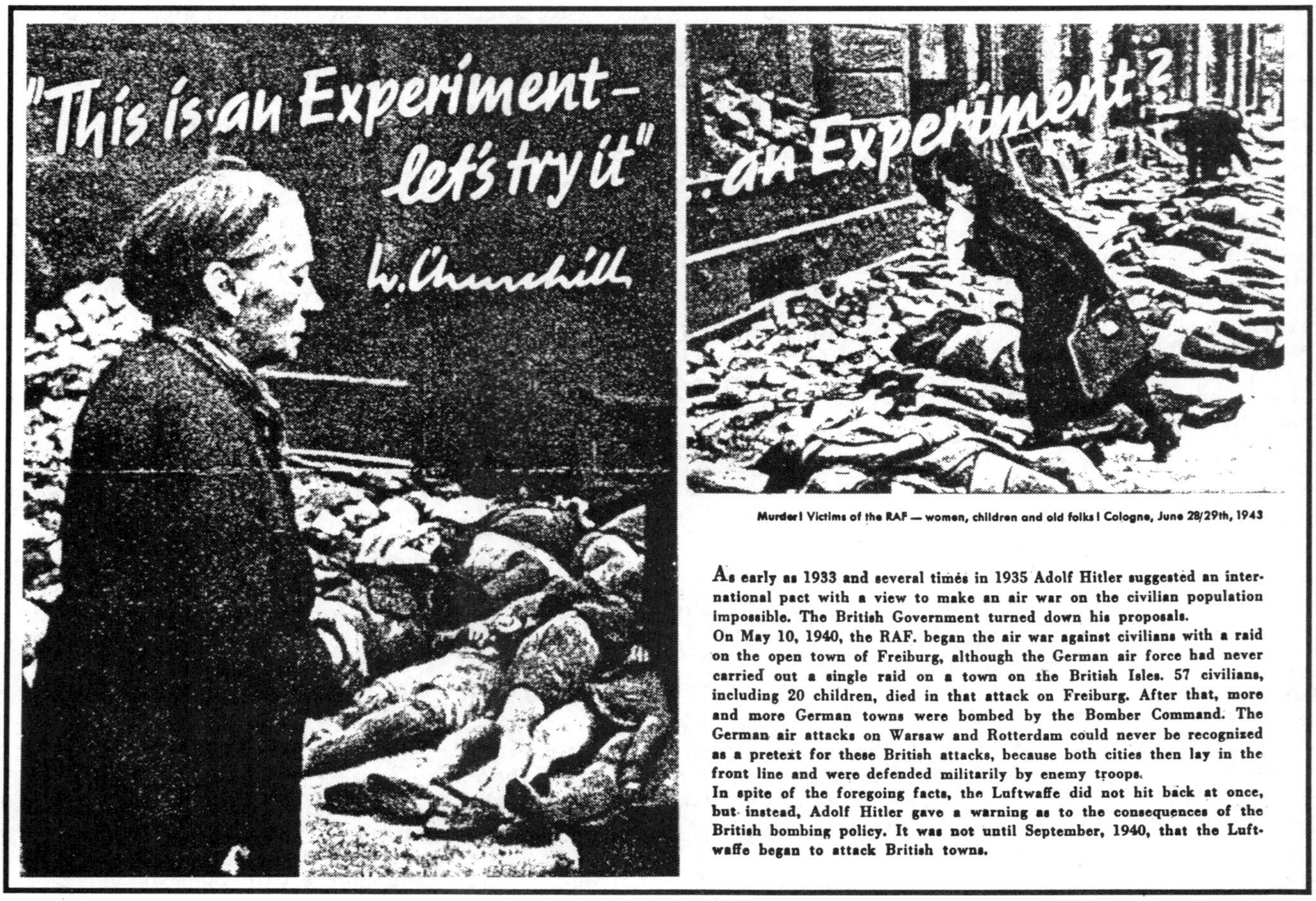

Murder! Victims of the RAF — women, children and old folks! Cologne, June 28/29th, 1943

As early as 1933 and several times in 1935 Adolf Hitler suggested an international pact with a view to make an air war on the civilian population impossible. The British Government turned down his proposals.
On May 10, 1940, the RAF. began the air war against civilians with a raid on the open town of Freiburg, although the German air force had never carried out a single raid on a town on the British Isles. 57 civilians, including 20 children, died in that attack on Freiburg. After that, more and more German towns were bombed by the Bomber Command. The German air attacks on Warsaw and Rotterdam could never be recognized as a pretext for these British attacks, because both cities then lay in the front line and were defended militarily by enemy troops.
In spite of the foregoing facts, the Luftwaffe did not hit back at once, but instead, Adolf Hitler gave a warning as to the consequences of the British bombing policy. It was not until September, 1940, that the Luftwaffe began to attack British towns.

Bacteriological Laboratory, to MI5, to the Home Office (Police Duty Room), to the Regional Information Officer and to GHQ Home Forces, the remaining copies being kept for three months in case required and then destroyed."

A clearance from the Medical Research Council, dated 23rd January 1945, reads:-

"Report upon Propaganda Leaflets said to have come from German sources and to have been scattered in different parts of England.

Leaflet headed 'This is an Experiment' - received 1st September 1944, from Hackney Police via Sector 6 Public Health Laboratory, St Mary's Hospital.

Leaflet headed, 'The Other Side' received November and December 1944.

Representative samplings of the above have been examined both by the culture method and animal inoculation. These tests did not disclose any evidence that the leaflets were contaminated with the germs of human or animal disease."

In the early morning of 23rd December, the day before the Manchester attack, two air-launched flying bombs crossed the South East coast, one to land at Oldbury, Kent, the other near Reading in Berkshire. Both carried two new types of German propaganda leaflet. These were to come in much greater numbers the following night.

One was a miniaturised version, printed in English, of the well-known "Signal" magazine. This was produced by the German army under the supervision of Dr Goebbels' Ministry of Propaganda. It was intended to advance the image and ideas of the Third Reich to the peoples of the occupied countries as well as to neutral states. "Signal" was published fortnightly from April 1940 to March 1945 and was somewhat similar in presentation to the American journal "Life", and to the British journal "Picture Post".

At the peak of its circulation it reached $2\frac{1}{2}$ million and it was translated into some 20 different languages. These included English for consumption in the Channel Islands, Ireland and, before December 1941, the United States.

The cost of all this was borne by the German High Command, who evidently considered it value for money as an influence in favour of their cause. After 1943 there was a reduction in the number of copies printed, but in 1944 "Signal" was made more attractive by the introduction of a full colour cover.

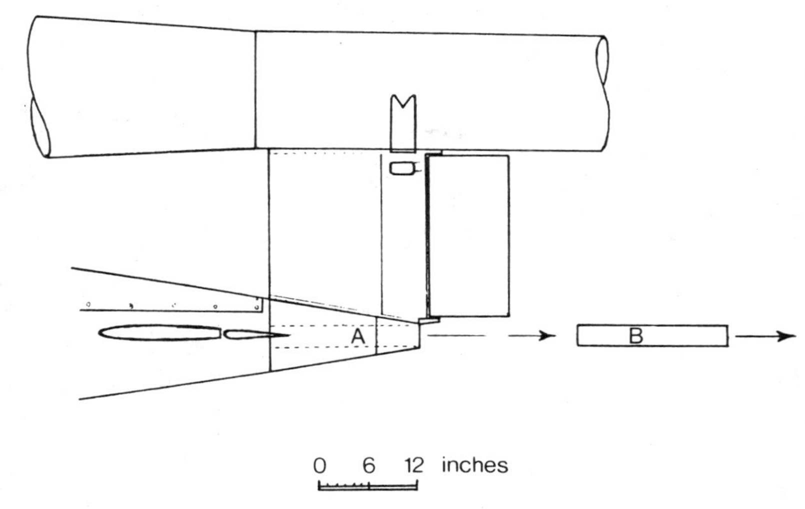

Leaflet-carrying flying bombs: 'A' is the paper cylinder, $2\frac{1}{2}$" by 18", positioned in the tail cone of the V-1. 'B' is the cylinder ejected from the tail cone at moment of motor cut-out.

The victorious image of the Third Reich, promoted in the early years of publication, gradually changed as the war news became less favourable, and by the second half of 1944 had become distinctly heavy and retrospective.

The miniature version delivered by V-1 was of Volume 5, No.17 of 1944, 8 x 4 inches in size. It still contained 38 pages within its covers, and all the content of the original full-size English language version. Naturally, in the miniaturised form the reduced size of the printing required the reader to have either extremely good eyesight or a powerful pair of spectacles. The coloured front cover showed a representation of a clock with hands at five minutes to twelve. In place of the figures on the clock face were the signs of the Zodiac. This introduced the theme of the particular issue, which was a depressing review of the European situation from the Nazi viewpoint, under two article headings, "Five to Twelve" and "Five past Twelve". A series of photographs under the title "Vulcan's Forge" purported to show the underground armaments factories from which, free from Allied bombing, Germany could keep the war going indefinitely.

Some coverage was given to lighter matters - the arts, attractive girls and a measure of advertising space.

The second leaflet, or rather series of leaflets, were single sheets, printed on both sides in black and red, on white paper, and headed "V-1 POW Post". At the top left hand corner of each leaflet was an oblong outline containing the words:- *"The finder is requested to cut out or copy the letters printed here and to transmit them to the addresses so that they receive them as early as possible. The original letters are being sent through the Red Cross in the usual mail channel."*

Two layouts were used, each 8 by 6 inches, one horizontal, the other vertical. Each leaflet had, at its bottom right hand corner, one of a series of numbers ranging from 1 to 6, each number preceded by a V. The number indicated to which of a series of six the particular leaflet belonged. Each contained three or four letters reproduced in facsimile to prove authenticity, each with a printed version alongside it. These purported to come from British prisoners of war in German hands. The addressee of each letter was clearly shown.

One typical letter was written by Raymond Joliffe, a Staff Sergeant glider pilot, who had been wounded and captured at the Arnhem landing. It was addressed to his wife, Cyrilla, in Whalley Range, Manchester. In 1969 a Manchester Evening News reporter handed a copy of the leaflet to Raymond Joliffe. It was, Mr Joliffe said, the first time he had seen the letter since he wrote it as a prisoner, and reading it brought back memories of "just how fed up" he was at the time. It said:-

"I got a burst of machine gun bullets in the left calf which just missed the bone but removed a large piece of meat. It is healing well. When my glider got burned I received burns to the right hand (now healed), right shoulder (now better), right arm (healing very well) and a large burn on the right leg, which was the worst. It is a large area and will take a long time to get well. I should be able to walk quite well when I am better, so do not worry..."

A note at the foot of the letter had been added by the German prison camp commandant, addressed to Mrs Joliffe:-

"You can rest assured that regarding the health of your husband, he gets all the attention possible."

Just before Christmas a German officer approached Staff Sergeant Joliffe in the hospital, and told him that he would be given the opportunity to send an extra letter through Red Cross channels and that this would reach its destination before Christmas Day, adding that it would be "going by air". So, indeed, it did, but not in the way envisaged by the writer.

The standing instructions to the police regarding seizure of enemy leaflets now assumed an extra importance. The Germans had a very clever ruse worked out for military use of the letters they talked the unsuspecting wounded prisoners into writing for this delivery of "V-1 POW Post".

Their hope was to learn where the leaflets, and therefore the bombs, had fallen, and from this to plot the accuracy of their missiles. The finder would, they hoped, send the letter on to the addressee, who would, again unsuspectingly, write back to the prisoner saying where the letter had been picked up. This would be the locality in which the V-1 had fallen, since the leaflets were not released until the missile made its final plunge. German censors of incoming POW mail would then pass the information to the Luftwaffe intelligence branch and hence to KG53.

Because of the standing order, and the fact that so many of the leaflets dropped were still bundled together, police recovered the vast majority. Even when the packages did eject and open, the still air of the early hours of a frosty December morning did not carry the leaflets very far. Once settled on the ground, they stayed where they were. Some were picked up and spirited away by souvenir hunters.

In order to prevent any which did escape detection from achieving their purpose, the addressees were discreetly approached by the police and warned not to reveal that they had received the letters concerned other than through the Red Cross Mail as normal. Two copies of Raymond Joliffe's letter, which were found in the East Didsbury or Gatley area, were delivered in good faith to Mrs Joliffe, in time for Christmas, by people who had found them. Another was delivered by the postman and had evidently come from the Oldham bomb.

A careful investigation into the letters was carried out and at first there was some doubt as to their authenticity. A report from the Kent County Constabulary, dated 1st January 1945, of investigations into the case of a leaflet picked up and posted from Yorkshire to Mr and Mrs Swift of Maidstone, told that "Mrs Swift has stated to neighbours that the writing is in no way similar to that of her son's, neither is the phraseology his". Mrs Swift, it seems, suspected a witch hunt in the offing.

The Ministry of Home Security wrote, on 27th December, in regard to "V-1 POW Post", that "So far only V1, 3, 4, 5 and 6 of this issue have reached the department. Sample checks carried out by the police of some of the addresses given in these alleged letters have proved that they are correct, but that the wording of some of the letters casts doubt upon their authenticity. It will also

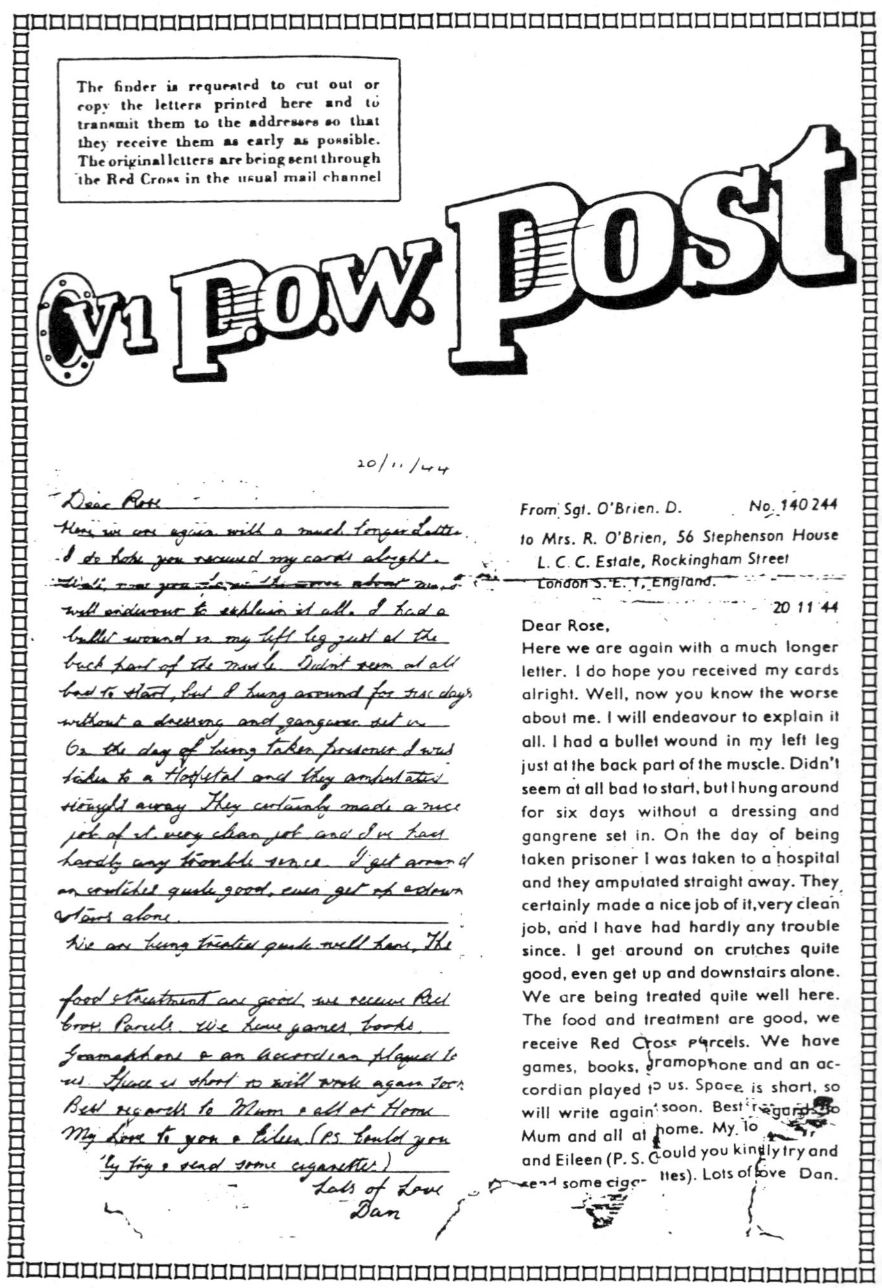

The finder is requested to cut out or copy the letters printed here and to transmit them to the addressees so that they receive them as early as possible. The original letters are being sent through the Red Cross in the usual mail channel

V1 P.O.W. Post

20/11/44

Dear Rose

Here we are again with a much longer letter. I do hope you received my cards alright. Well, now you know the worse about me I will endeavour to explain it all. I had a bullet wound in my left leg just at the back part of the muscle. Didn't seem at all bad to start, but I hung around for six days without a dressing and gangrene set in. On the day of being taken prisoner I was taken to a Hospital and they amputated straight away. They certainly made a nice job of it very clean job and I've had hardly any trouble since. I get around on crutches quite good, even get up and down stairs alone.

We are being treated quite well here. The food & treatment are good, we receive Red Cross Parcels. We have games, books, Gramaphone & an accordian played to us. Space is short so will write again soon. Best regards to Mum & all at Home My Love to you & Eileen (P.S. Could you try & send some cigarettes)

Lots of Love
Dan

From Sgt. O'Brien. D. No. 140244
to Mrs. R. O'Brien, 56 Stephenson House
L. C. C. Estate, Rockingham Street
London S. E. 1, England.

20 11 44

Dear Rose,
Here we are again with a much longer letter. I do hope you received my cards alright. Well, now you know the worse about me. I will endeavour to explain it all. I had a bullet wound in my left leg just at the back part of the muscle. Didn't seem at all bad to start, but I hung around for six days without a dressing and gangrene set in. On the day of being taken prisoner I was taken to a hospital and they amputated straight away. They certainly made a nice job of it, very clean job, and I have had hardly any trouble since. I get around on crutches quite good, even get up and downstairs alone. We are being treated quite well here. The food and treatment are good, we receive Red Cross parcels. We have games, books, gramophone and an accordian played to us. Space is short, so will write again soon. Best regards to Mum and all at home. My love to you and Eileen (P. S. Could you kindly try and send some cigarettes). Lots of love Dan.

A leaflet which fell from the V-1 at Kelsall, Cheshire.

be seen that several errors in transcription occur in the printed version."

V3 was the leaflet of the series most widely distributed during the Manchester raid and it was identified as having fallen at Turton, Oswaldtwistle, Oldham, Tottington, Kelsall, Grange Moor, Rossington and Willerby. It contained three letters and is reproduced here.

Each letter usually had an encouraging postscript from either the British Medical Officer or from the German camp commandant. It is difficult to resist the conclusion that the Germans had selected those letters which tended to show them in the best light as captors, a propaganda bonus objective.

V4 appeared at Brindle, Ollerton and Didsbury; V5 at Didsbury and South Cliffe; V6 at Radcliffe, Worsley and Adswood. "Signal" magazine was found at Sturton-le-Steeple, Newport, Kelsall, Macclesfield Forest, Rossington, Epworth and Woodford. At Epworth only two copies were reported found by the ROC observers, who were the first to venture to the crater. These were still in a paper cylinder, but what happened to the rest of the bundle is not clear.

The small scale of the leaflet raid, indeed, of German leaflet raids generally on the United Kingdom, is brought into perspective when it is remembered that 6,500,000,000 Allied propaganda leaflets were dropped over the Axis homelands and occupied territories in the European theatre during the war. The flying bomb leaflet-dropping exercise would not appear to have achieved either of its objectives in the slightest degree.

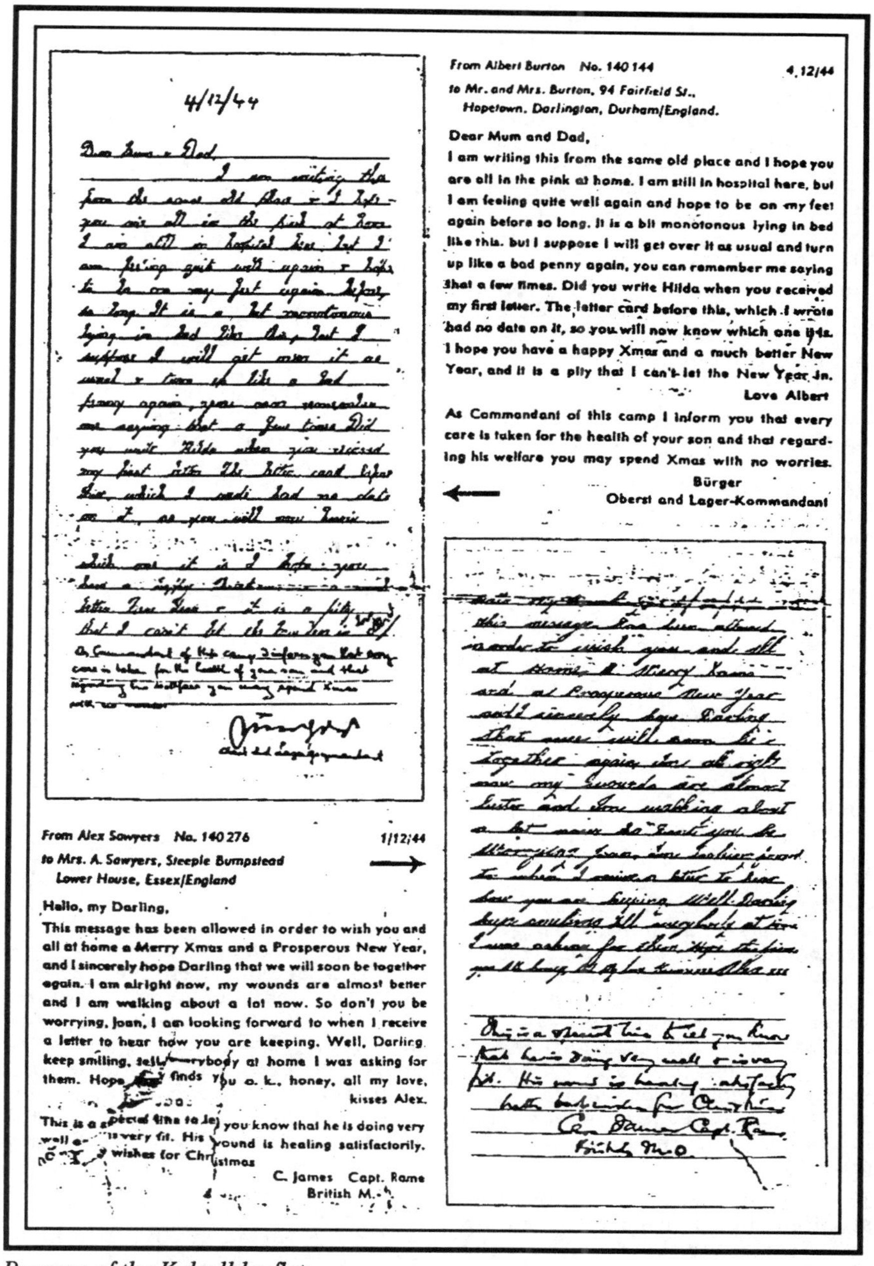

From Albert Burton No. 140 144 4.12/44
to Mr. and Mrs. Burton, 94 Fairfield St., Hopetown, Darlington, Durham/England.

Dear Mum and Dad,
I am writing this from the same old place and I hope you are all in the pink at home. I am still in hospital here, but I am feeling quite well again and hope to be on my feet again before so long. It is a bit monotonous lying in bed like this, but I suppose I will get over it as usual and turn up like a bad penny again, you can remember me saying that a few times. Did you write Hilda when you received my first letter. The letter card before this, which I wrote had no date on it, so you will now know which one this. I hope you have a happy Xmas and a much better New Year, and it is a pity that I can't let the New Year in.
Love Albert

As Commandant of this camp I inform you that every care is taken for the health of your son and that regarding his welfare you may spend Xmas with no worries.
Bürger
Oberst and Lager-Kommandant

From Alex Sawyers No. 140 276 1/12/44
to Mrs. A. Sawyers, Steeple Bumpstead Lower House, Essex/England

Hello, my Darling,
This message has been allowed in order to wish you and all at home a Merry Xmas and a Prosperous New Year, and I sincerely hope Darling that we will soon be together again. I am alright now, my wounds are almost better and I am walking about a lot now. So don't you be worrying, Joan, I am looking forward to when I receive a letter to hear how you are keeping. Well, Darling keep smiling, tell everybody at home I was asking for them. Hope [illegible] finds you o. k., honey, all my love,
kisses Alex.

This is a special line to let you know that he is doing very well [illegible] is very fit. His wound is healing satisfactorily. [illegible] wishes for Christmas
C. James Capt. Rame
British M.O.

Reverse of the Kelsall leaflet

Chapter 11

'WHICH MAY WELL BE REPEATED'

Kampfgeschwader 53 had sprung a surprise by directing an attack at Manchester to the extent that there had been no prior intelligence of the intended raid and no actual defensive deployment to meet it.

The anticipatory plan of No.5 Anti-Aircraft Group was already out of date when orders were received for the limited move into the Fringe on the morning of 24th December. The orders came at a time of desperate manpower shortage in Anti-Aircraft Command. Successive withdrawals of men from it had been made in the autumn and early winter to fulfil the needs of the land forces on the Continent, and particularly of the anti-aircraft artillery units covering Antwerp, Brussels and Liege, against which a severe ramp-launched V-1 attack was opened.

Early in December, in order to meet yet another clamour for men to reinforce the ever-needful army on the frontier of Germany, the War Office began to press unrelentingly for every fit man under 35 years of age in UK Anti-Aircraft Command to be transferred to the infantry. The prospect for AA Command was of a new cut of 40,000 men, many of them its best and most skilled gunners. Some disbandment of units and a loss of efficiency was inevitable.

Partially stood down, the Civil Defence Services in the North West seemed less than adequate to meet even a sporadic air offensive, despite the valiant efforts of individuals on the day following the raid. Few of the full-time Civil Defence Personnel remained in service. The part-timers were still committed, but lacking actual air raid experience for $2\frac{1}{2}$ years and demoralised by the stand down, they were hardly in the best shape to meet such a threat.

A meeting was called of all the heads of the Civil Defence services in the North West Region, at which Hartley Shawcross clearly set out the lessons to be learned by the Civil Defence organisation from the flying bomb raid, particularly as a result of the Oldham incident. All Civil Defence workers were to be called together to hear a lecture on these lessons.

On 17th January, the Minister of Home Security issued a warning letter to all Civil Defence Regions headed, "Air Attacks on Areas which have had an immunity for some time". It betrayed some official alarm:-

"The recent flying bomb attack on the North, which may well be repeated and may be extended to other areas, has emphasised the importance of ensuring that the Civil Defence Services are kept up on a state of operational efficiency... Many Wardens' Posts have been shut, and in some areas, Wardens' Posts are only open for a short period in the early evening... Many Depots are now only manned on the siren sounding or enemy action occurring. Telephones should be tested frequently at these posts to be used for operations... All personnel must be definitely informed of their obligations to turn out on an alert, or on enemy action occurring... and of the consequences which may follow a failure to report for duty."

The letter was probably instigated by a report drawn up by Hartley Shawcross on the attack. This was in some respects critical of the way in which the Oldham incident had been handled. He commented that an incident of that size was too large to be controlled by one Incident Officer and that the organisation generally needed an overhaul.

Despite the warnings, hope was allowed to triumph over caution. Apart from minor moves, such as the provision of rescue vehicles at Hyde, further resources were not committed to Civil Defence activity. The releases and discharges of personnel continued. The total number of full-time paid Civil Defence staff throughout the country, which stood at 217,742 at the end of September 1944, was down to 146,467 at the end of January 1945 and continued to decline rapidly after that.

Formal training had virtually ceased in Manchester by the middle of December, owing mainly to the discharge of the full-time personnel, and was not restarted. The same applied generally throughout the North West Region.

The feeling was abroad that the end of the war in Europe could not now be long delayed and there was a disinclination to take air raids seriously any more.

The Civil Defence log books for the Manchester area do record one further air raid alarm as due to flying bomb activity. Typically, that held in the Rochdale Local History Library, "German Air Raids and Warnings on Rochdale", records:-

Friday 29th December 1944.
Flying Bombs.
Early Morning. Wailer 8.15am
27 minutes Steady Blast 8.42am

There was no second flying bomb raid; the cause of the warning was mis-stated to the Civil Defence Authorities. The true cause of the alert of 29th December was an American Liberator bomber with bombs on board, which ran into trouble over the area and was considered to be of sufficient danger to justify a public warning.

In retrospect, from a military viewpoint, the strain and risk to KG53 of an offensive against targets in Northern England and the Midlands would appear justified. Dangerous and difficult though such air-launch operations at long range may have been, opportunity did exist, in the late summer and autumn of 1944, for the Luftwaffe to mount an extremely cost-effective series of diversionary V-1 attacks on the Midlands and Northern England industrial areas. The threat to such vital centres of war production would have forced a major diversion of resources to defensive effort and still left the initiative in the hands of the attacker. Nor can the internal propaganda value of such an offensive be ignored.

Instead, a dogged, irregular pounding away at London continued in the face of a strong, concentrated and efficient defence, able to shoot down 63.2% of the 638 flying bombs which approached the East Coast, aimed at the city, between 16th September 1944 and 14th January 1945. From the 638, only 67 impacted in the urban area of London.

Once it was decided to place operational control of V-1 wholly in the hands of the army Commander in Chief West, it was only natural that the tactical thinking of the army should prevail. The prime V-weapon of the army available for the bombardment of Britain was the V-2 rocket. After this came into use on 8th September, the limited air-launch effort which the Luftwaffe was able to provide was used by the army as a back-up to its V-2 offensive against London. This misuse of the flexibility of V-1 air-launching was the indirect result of the generally unhappy mix of inter-service rivalry, personality conflict and political interference which had such a profound effect on much of the German High Command's planning. It was reflected on this occasion in the lack of imagination shown by Colonel Eugen Walther and 65 Corps headquarters.

Meanwhile, the crippling consequences of the increasing aviation fuel shortage, more

This Heinkel III H-16, Works No.161487, of 3/KG53 stands in front of an improvised hangar at Vechta Airfield with V-1 in launching position below the starboard wing root